• Volcanoes in Human History

Volcanoes
in Human History

The Far-Reaching
Effects of Major
Eruptions

Jelle Zeilinga de Boer

and

Donald Theodore Sanders

PRINCETON UNIVERSITY PRESS

Princeton and Oxford

Published by Princeton University Press, 41 William Street,
Princeton, New Jersey 08540

In the United Kingdom: Princeton University Press, 3 Market Place,
Woodstock, Oxfordshire OX20 1SY

Library of Congress Control Number 2001095818

ISBN 0-691-05081-3 (hardcover : alk. paper)

British Library Cataloging-in-Publication Data is available

This book has been composed in Palatino, Fairfield, and Copperplate
by Princeton Editorial Associates, Inc., Scottsdale, Arizona

Printed on acid-free paper. ∞

www.pup.princeton.edu

Printed in the United States of America

1 3 5 7 9 10 8 6 4 2

The following publishers have generously given permission to use quotations from copyrighted works. From "The Control of Nature," by John McPhee. Reprinted by permission; copyright © 1988 by John McPhee. Originally published in *The New Yorker*. All rights reserved. From *The Letters of the Younger Pliny*, translated by Betty Radice (Penguin Classics, 1963), copyright © Betty Radice, 1963, 1969. Reproduced by permission of Penguin Books Ltd. From *Volcanoes of the Earth*, Revised Edition, by Fred M. Bullard, copyright © 1976. By permission of the University of Texas Press. From *Surtsey* by Sigurdur Thorarinsson, copyright © 1964, 1966 by Almenna Bokafelagid. Used by permission of Viking Penguin, a division of Penguin Putnam Inc. From "Dynamic Mixing of Water and Lava," by S. A. Colgate and Thorbjörn Sigurgeirsson. Reprinted by permission of *Nature* (244, August 31, 1973: 552), copyright © 1973 Macmillan Magazines Ltd.

To Joe Webb Peoples—
teacher, mentor, and friend to both authors

• Contents

· Foreword

MOST PEOPLE seldom think about volcanoes or the role they have played in human history. That is because most of us do not live where volcanoes are erupting. They are not part of our everyday lives.

But if you lived near Mount St. Helens when it exploded in 1980, you will not soon forget its tremendous eruptive power. In Iceland, which sits astride the Mid-Atlantic Ridge, volcanoes dominate people's lives and their mythology. Those who live on the Icelandic island of Heimaey literally have an active volcano in their backyards. Residents of Reykjavik, the country's capital, have their homes heated by water that circulates through hot lava.

Active volcanoes remind us that the earth is a living, breathing organism, making our planet unique compared with its sister planets Mercury, Venus, and Mars. In addition to having an immediate effect upon those of us who live near them, volcanoes, in the long term, can profoundly affect our very psyches.

The authors of this book, Jelle Zeilinga de Boer and Donald Theodore Sanders, examine the relationship between volcanoes and human history through nine case histories. These chilling examples show the profound impact volcanic eruptions have had upon humans. The incredible story has taken centuries to unfold. It will surely continue to evolve.

Robert D. Ballard
Institute for Exploration, Mystic, Connecticut

• Preface

THERE IS A WIDESPREAD PERCEPTION that the sciences and the humanities are incompatible, that they have little or nothing in common. What do history, the arts, and great literature have to do with physics, chemistry, biology—or earth science? In 1959 the British scholar C. P. Snow analyzed that question in his widely read book *The Two Cultures*.[1] Snow attributed the apparent incompatibility to misinterpretation and lack of understanding on both sides, and in his book he attempts to reconcile the "two cultures."

The notion that Snow's two cultures are at odds is trenchantly expressed in a novel published in 1983 by the American author Trevanian in a scene where one of the characters warns another, "Beware the attraction of the *pure* sciences. They are pure only in the way an ancient nun is—bloodless, without passion. No, no. Stick to the humanistic studies where, though the truth is more difficult to establish and the proofs are more fragile, yet there is the breath of living man in them."[2]

One of the present authors (Zeilinga de Boer) has attempted to bring the two cultures together at Wesleyan University in his course on geological catastrophes, in which he demonstrates to liberal-arts students that the sciences are not "bloodless," that in the earth sciences in particular, something akin to the "breath of living man" can be seen in such phenomena as volcanic eruptions and earthquakes. In his lectures, Zeilinga de Boer discusses selected geological events, describing their origins while emphasizing the many ways in which the events have affected people, societies, cultures, even history itself. This book, which grew out of those lectures, has as its theme the human dimension of volcanism. The ways in which earthquakes and the humanities are intertwined will be discussed in a later volume.

Volcanic eruptions are treated only descriptively in most books, the descriptions concerned mainly with environmental consequences and the number of casualties. Many of these short-lived events, however, have had long-lasting aftereffects. Some eruptions have had global consequences that have lasted for years, decades, centuries, even millennia. Some of the events can be described as catalytic in the sense that their direct aftereffects give rise to other phenomena, whether environmental, economic, or cultural.

Moreover, most books treat only the destructive side of volcanism. But volcanic eruptions, devastating as they may be in the short term, have been of long-lasting benefit to humankind in many ways. Volcanic soils are among the most fertile on earth. Many of our mineral resources are of volcanic origin. Even water, the basic resource without which life could not exist on our planet, ultimately is created by volcanic activity within the earth. All these aspects of volcanism—destructive as well as beneficial—are discussed in the chapters that follow.

In this book we explore nine volcanic eruptions. In each case we briefly discuss the geological setting in terms of plate tectonics—the theory that virtually rigid segments of the earth's crust move about over a less rigid layer and collide, and that the collisions give rise to earthquakes and volcanic activity. Then we discuss the aftereffects of the eruption—its consequences—in human terms.

By describing not only the immediate physical effects of volcanism but also the long-term aftereffects, we demonstrate the inherent connections that exist between the earth sciences and the humanities. Some eruptions have changed societies. Some have been followed by famine and disease, or by political changes either peaceful or violent. Others, of truly ancient origin, have passed into mythology or are reflected today in religious beliefs and practices. Some have achieved cultural immortality in the arts or in literature—in paintings, poems, great books, operas, motion pictures, even architecture.

Volcanism is the surface manifestation of a living earth. We can think of a volcanic eruption as the plucking of a long, tight-stretched string representing time: when the string is plucked it vibrates. During the eruption, at the point of origin, where a great deal of energy is being released, the vibrations will have high amplitudes and short wavelengths. The vibrations will be powerful, but each will last only a moment. Farther along on the string, with the passage of time, the amplitudes will decrease and the wavelengths increase. That is to say, the aftereffects will become less intense and they will last longer, as shown in the figure below.

The "vibrating string" showing the long duration of interdisciplinary effects that can follow a volcanic eruption.

For example in 1815, Tambora, a volcano in Indonesia, exploded in the greatest eruption known to history. It killed perhaps 70,000 people outright. Regionally the catastrophe devastated forests and croplands, producing famine and disease. Around the world there were major changes in the weather as dust and aerosols from the eruption, carried by high-altitude winds, circled the globe and dimmed the sun's rays. In Europe, prolonged inclement weather caused crop failures and food riots, and in 1816 North America suffered the infamous "year without a summer." The European weather inspired Lord Byron's gloomy poem "Darkness" and Mary Shelley's immortal novel *Frankenstein,* which continues to attract readers and moviegoers almost two centuries later. Tambora's string vibrates to this day.

By discussing Tambora's "vibrating string," and eight others, we hope to draw interest both to the tectonic origin of specific volcanic eruptions and to their interdisciplinary consequences. When most of those eruptions occurred, the earth was sparsely populated. Today the human population exceeds 6 billion. The geological events discussed here are not unique. Similar events will occur in the future, and their effects will be magnified by the population density of our crowded planet. It is crucial that we understand the origin of volcanism as well as the devastation it can cause, and the aftereffects, for good or ill, that can linger for years, even decades, to come.

· Acknowledgments

THE AUTHORS GRATEFULLY ACKNOWLEDGE the keen editorial assistance of Kristin Gager and Joe Wisnovsky of Princeton University Press, as well as the thoughtful professional reviews of Susan Hough, Michael Ort, and others in developing the final manuscript. We thank James Gutmann, Gerrit Lekkerkerker, Johan Varekamp, Alison Hart, and Mary Watson for their many helpful suggestions and their encouragement throughout the writing of the book. We also thank Gordon Eaton for his comprehensive review of our chapter on volcanism in Hawaii. Not least, we extend our thanks to research librarians at Olin Library, the Science Library, and the Art Library at Wesleyan University, as well as in the town libraries of Fairfield and Madison, Connecticut. They all generously helped us locate information that was important to our work.

Special thanks go to Michael Ross, whose telephone call in 1994 led, more or less directly, to the authors' collaboration. And of course we thank Edward Knappman of New England Publishing Associates, who agreed to represent us, provided invaluable guidance, and on our behalf contacted Princeton University Press.

More personally we wish to thank Felicité de Boer, Joan Boutelle, and Katherine Sanders for their support and help in so many ways.

· Table of Conversions

1 centimeter = 0.39 inch
1 meter = 3.28 feet
1 kilometer = 0.62 mile
1 square meter = 10.76 square feet
1 square kilometer = 0.39 square mile
1 hectare = 2.47 acres
1 cubic meter = 35.31 cubic feet
1 cubic kilometer = 0.24 cubic mile
$9/5 \times$ degrees Celsius + 32 = degrees Fahrenheit

• Volcanoes in Human History

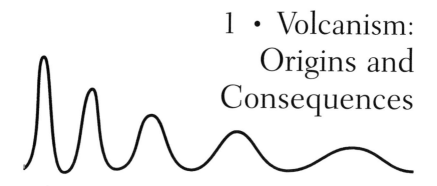

1 · Volcanism: Origins and Consequences

Giant smoking volcanoes
stand in a row
like the pipes of a cosmic organ
through which the mighty breath of the earth
blows its roaring music

Robert Scholten

WHEN OUR ANCESTORS realized that their world was not a flat disk resting on the back of a giant turtle—that instead, the earth is a spheroid whirling through space in orbit around the sun—they began to comprehend the nature of the planet that is our home. Over many centuries, scientists pieced together a great deal of information about the earth—the materials of which it is composed, the atmosphere surrounding it, the infinite variety of landforms on its surface, the kinds of rocks that are exposed there.

Eventually, by studying earthquake waves and the time they take to pass through the earth, scientists deduced that our planet has a dense, at least partly molten core at its center and that the core is overlain by a thick layer of less dense material, which they named the mantle. Above the mantle is the thin, rocky crust upon which we live. We might say that the earth resembles an apple in some respects. If an apple is sliced in two, the cross section reveals a small, circular "core" (where the seeds are), a thick "mantle" (the edible flesh), and a "crust" (the very thin skin). The relative proportions of those parts of an apple are not unlike the proportions of the main parts of the earth.

Like our understanding of the structure of the earth, our understanding of volcanoes slowly emerged from beliefs conceived in ignorance. Well into the European Middle Ages, many people thought of volcanoes, with their fiery summits and unearthly roarings, as entrances to the underworld, the hellish world of suffering sinners. In the early 1300s, the Italian poet Dante Alighieri captured the prevailing views of that time in his masterpiece, the *Divine Comedy,* an allegorical, three-part portrayal of a journey, first into hell, the realm of eternal punishment, then into purgatory, where there is hope for the soul's salvation, and ultimately into paradise, where the soul returns to God. Dante's hell is a fiery cavity that reaches to the center of the earth, where the devil dwells. What more obvious interconnection could there be between the devil's subterranean realm and the external world of the living than a volcano?

Forces of destruction, sources of bounty

With the maturing of the geological sciences, of course, such beliefs faded into fantasy. But the association of volcanoes with suffering and disaster remained, for volcanoes, after all, can be, and often are, deadly and destructive.* During the past 400 years, perhaps a quarter of a million people have been killed as a direct result of volcanic eruptions. Indirect aftereffects, such as famine and disease, may well have tripled that number.

---

*The term *volcano* can be defined in different ways. The dictionary definition includes any opening in the earth's crust through which molten lava, volcanic ash, and gases are ejected. The term can also refer to a mountain formed by the materials ejected from such an opening. Strictly, then, a volcano can be anything from a vent or fissure in the earth to a mountain with a height measured in kilometers. In this book, for simplicity, we reserve the term for volcanic mountains.

Volcanic lava flows consume everything in their path. Volcanoes also can cause landslides and mudflows that rapidly travel long distances, wreaking havoc. Volcanic dust and aerosols in the atmosphere can shield the earth from sunlight and the sun's warmth, disastrously altering weather patterns, sometimes for years. French poet Max Gérard eloquently sums up this calamitous side of volcanism:

> Here is Wotan's brazier,
> Vulcan's furnace,
> the forge of Cyclops,
> Satan's pyre!
> Here is the first panting,
> the birth of matter,
> here the Gods are stoking
> the superstition of men,
> here the times are coming
> of violence and damnation![1]

But paradoxically there are many beneficial aspects of volcanism, and they are crucially important to our lives. Over the eons, volcanic eruptions have emitted vast amounts of water vapor, bringing to the surface the fluid that is essential to life. Much of the water vapor in any given eruption may come from volcanically heated groundwater—recycled rain and snow in the zone of saturation below the surface of the ground. But many scientists believe that all the water on earth—whether in clouds, mountain streams, rivers, lakes, or oceans—was originally vented into the atmosphere by volcanoes. According to that theory, water originated as dissociated hydrogen and oxygen atoms deep in the earth's mantle. Volcanism is responsible, too, for creating many of the minerals in the earth—minerals in the ores that give us copper, lead, zinc, and other metals required for industry and modern technology.

Volcanic eruptions also bring nutrients to the earth's soils. The potassium and phosphorus needed by plants are contained in the ash produced by many eruptions. The weathering of volcanic rocks also releases such nutrients. Therefore volcanism supports plant life and is ultimately responsible, in many regions, for agricultural abundance. Hundreds of millions of people live quietly on the flanks of volcanoes or in nearby low-lands, farming the fertile soils. Thus, though volcanoes are destructive during short periods of eruption, they bring us many essential benefits during the long periods between erup-tions. This all-important, and often neglected, dual view of volcanism is vividly illustrated in Figure 1-1, which shows a volcano erupting and bringing death and destruction while, at the same time, producing a cornucopia overflowing with the good things of life. Again quoting Max Gérard,

> It burns so as to re-create,
> the glow of fire becomes an embrace . . .
> that destroys and rebuilds, tears and will mend, burns
> and will make green again.[2]

Products of volcanism

The products of volcanic eruptions—lava, gases, and frag-mental materials such as ash—all ultimately derive from molten rock, called *magma*, that originates within the earth. Because magma is hot and fluid and contains dissolved gases, it is less dense than solid rock and tends to work its way upward through fissures in the earth's crust. Lava is magma that has erupted at the surface. The term *lava* applies both to the molten material and to the rock that forms after magma has cooled and hardened. Rapid cooling, which leaves little time for mineral crystals to form, produces fine-grained rock.

We often think of volcanic rocks as being black, or at least dark gray, creating dismal, colorless landscapes. Most lava flows are indeed drab and dark, but some, depending on their chemical composition, create landscapes that are vibrant with color. In 1924 after Gilbert Grosvenor, a founder of the National Geographic Society, climbed Mauna Loa, the largest volcano on the island of Hawaii, he reported traversing "a lumpy, rolling sheet of colored glass, extending as far as the eye could reach, glistening at times with the radiance of countless jewels,

FIGURE 1-1. The dual nature of volcanism. Volcanic eruptions cause death and destruction. But equally important in the long run, they provide fertile soils, hence bountiful harvests, as well as a wide range of mineral resources. Engraving by Nicollet after a design by Fragonard. Private collection.

sparkling with the brilliance of diamonds and rubies and sapphires or softly glowing like black opals and iridescent pearls."[3]

The gases released in volcanic eruptions comprise mostly water vapor, along with lesser volumes of carbon dioxide, sulfur dioxide, and other gases. Indeed, it is thought that volcanism was responsible for creating the planet's atmosphere when the earth was young. The oxygen we breathe came later, after the evolution of life-forms capable of photosynthesis, which uses sunlight to transform carbon dioxide and water into organic matter, releasing oxygen as a by-product.

Many of the materials ejected during eruptions are fragments of rock, either solidified bits of magma or pieces of preexisting rock torn from the conduit that feeds the volcano. Such materials are called *pyroclastic,* from the Greek *pyro* (fire) and *klastos* (broken). Sometimes clouds of such fragmental material, along with hot volcanic gases, form devastating pyroclastic flows, which, because of their weight, hug the ground and race down mountainsides at express-train speed, destroying everything in their path. Typically they separate into three parts:

- Dense material—fragments of fresh magma, pumice, and older volcanic rock ripped from the conduit or from the flanks of the volcano—that hugs the ground.

- Fiery, gaseous surges containing droplets of fresh magma. Many surges form at the head of the flow or along its sides, and they move much faster than the dense material.

- Clouds of volcanic dust that form buoyant plumes rising thousands of meters into the air.

Life cycles of volcanoes

Volcanoes have life cycles much as animals and plants do. On the morning of February 20, 1943, a Mexican farmer named Dionisio Pulido had the unpleasant experience of witnessing the birth of a volcano in his cornfield, about 320 kilometers west of Mexico City. What had been a slight depression in the field became a gaping fissure that emitted clouds of sulfurous smoke accompanied by loud hissing noises. By the next morning, Señor Pulido's cornfield was occupied by a cinder cone more than 10 meters high. Within a week the volcano, named Paricutín after a nearby village, had attained a height of 170 meters, and within a year it had reached 370 meters. Within nine years, Paricutín had produced voluminous lava flows that destroyed several towns and had grown to an elevation of 2,272 meters. Then the volcano went into repose.

In 1980 the Japanese author Shusaku Endo wrote a novel entitled *Volcano* in which the protagonist recalls how a university professor, Dr. Koriyama, eloquently described such a cycle: "A volcano resembles human life. In youth it gives rein to passions, and burns with fire. It spurts out lava. But when it has grown old, it assumes the burden of past evil deeds, and it turns quiet as a grave."[4] The fictional Dr. Koriyama might well have added that upon aging, volcanoes also lose much of their beauty. Young volcanoes typically form sleek, symmetrical cones. Old volcanoes have ragged, time-worn summits and flanks scarred by erosion.

Volcanoes erupt spasmodically, each eruption possibly including several pulses. Such activity can last from a few weeks to several years. Some volcanoes become quiescent, or dormant, for hundreds or even thousands of years but then are reactivated when a new upwelling of magma rises through the volcano's conduit. But all volcanoes eventually grow old and "die," or become extinct. Most have short life spans in geological terms—only one or two million years, often less. Volcanic fissures typically have even shorter life spans. Some of the magma that fills a fissure inevitably cools and solidifies there, forming a tabular body of rock called a *dike*. Any new pulses of magma normally intrude along a margin of the dike or through new fissures adjacent to it.

Volcanoes typically are crowned by eruption craters. During the largest eruptions, however, molten rock may not be able to rise from within the earth fast enough to replace the ejected magma, and as a result, the upper part of the volcano collapses inward. The result is not just a crater but a much larger depression called a *caldera* (Spanish for *caldron*): some calderas can be tens of kilometers in diameter. An example is the misnamed Crater Lake in southwestern Oregon. The lake occupies a caldera (not a crater) almost 10 kilometers across and about 600 meters deep. It was created about 6,000 years ago, when an ancient volcano known as Mount Mazama exploded.

Within Crater Lake lies Wizard Island, a small volcano, now extinct, that was born sometime after the caldera was formed—evidence that even apparently "dead" volcanoes can be reborn. A recent example of such rebirth occurred in 1927, when a volcano named Anak Krakatau appeared in the Sunda Strait between Java and Sumatra. Its birthplace was a submerged caldera that had been formed in 1883, when a volcanic island named Krakatau exploded in one of the great eruptions of history. Fittingly, the Indonesian name Anak Krakatau means "child of Krakatau."

Plate tectonics

In the 1960s geologists began to understand that the outer part of the earth is made up of individual rigid plates, some very large, others small, which slowly move over a ductile, or plastic, interior layer (Figure 1-2). The movement of these tectonic (structural) plates, at a rate typically measured in centimeters per year, is responsible for most volcanoes and earthquakes. This is the theory of plate tectonics, which revolutionized the science of geology by providing a single, unifying concept that helps explain most geological processes and features.

The earth's rigid outer shell includes the rocky crust and a thin layer of the uppermost part of the mantle. Together they form what geologists call the *lithosphere,* from the Greek *lithos* (stone). The ductile layer of mantle material over which segments of the lithosphere move is called the *asthenosphere,* from the Greek *asthenos* (weak).

The lithosphere segments—that is, the tectonic plates—are in motion presumably because of slowly moving convection currents within the mantle. The currents are believed to be driven by heat from the earth's core, much as convection currents are created in a pot of water heated on a stove. Hot water, being less dense than cold water, rises to the surface, where it cools, becomes more dense, and therefore returns to

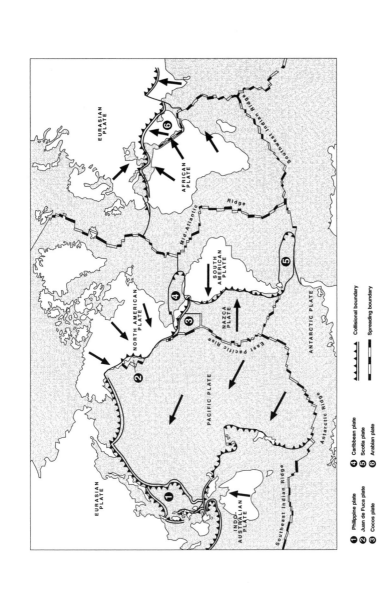

FIGURE 1-2. Configuration of the earth's tectonic plates, showing the collisional boundaries between converging plates and the spreading boundaries between diverging plates. The arrows indicate the present directions of plate motion. The black triangles at the collisional boundaries show the direction in which one plate is being subducted beneath another.

the bottom of the pot. A similar process is believed to be at work, albeit very slowly, within the earth.

As tectonic plates move about the earth's surface, inevitably they collide with one another. When they do, the consequences are profound. At these collisional, or convergent, boundaries, one plate slides beneath the other in a process known as *subduction*. The subducted plate descends into the asthenosphere, where high temperatures and pressures force fluids out of the subducted rock. The hot fluids—mostly steam from water in fractures and from minerals containing hydroxyl groups (comprising one hydrogen atom and one oxygen atom bound together)—rise and react with the rock in the wedge of mantle material above the subducted plate, causing chemical changes that locally reduce melting temperatures (see Figure 1-3). As a result, part of the asthenosphere wedge melts and becomes magma.

Magma formation

Volatile gases are released from the subducted plate as it reaches a depth of about 70 kilometers. By the time it has descended to 200 kilometers all liquids and gases have been squeezed out. Therefore it is between 100 and 150 kilometers that magma is generated. Blobs of magma are believed to rise slowly through the ductile asthenosphere, like air bubbles rising through water, until they reach the bottom of the solid lithosphere above the mantle wedge. There they coalesce into sheets of molten material that is hot enough to melt adjacent parts of the lithosphere.

As new batches of magma arrive, the molten mass eventually generates enough pressure to arch the still-brittle part of the lithosphere above it. Arching of the lithosphere creates fractures that allow magma to rise into the crust, where it forms pockets called *magma chambers* that may have volumes of many cubic kilometers. These chambers expand as more magma rises into them and as the hot magma melts rock formations that enclose them. As long as magma

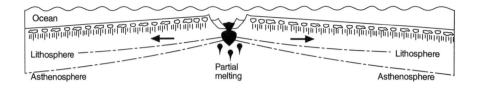

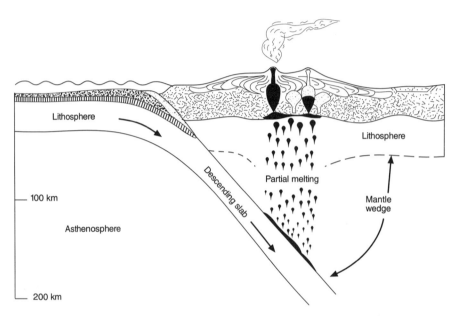

FIGURE 1-3. *Top:* Partial melting of the upper asthenosphere and formation of magma below an oceanic ridge at the boundary between separating plates. *Bottom:* Subduction of an oceanic plate beneath a continental plate and partial melting of the upper asthenosphere.

resides in a chamber, it continues to react with the surrounding rock, and its chemistry changes. It becomes lighter, less dense, and richer in gases, and it also becomes more viscous, or resistant to flow. Magma chambers give rise to volcanoes when increasing pressures force part of the molten mass up through crustal fractures, or conduits, that reach the earth's surface.

Magma can contain as much as 5 percent water by weight. Although not high in absolute terms, such a percen-

tage means that those huge subterranean magma chambers contain enormous quantities of water.* When the molten rock erupts at the surface, the hot, vaporized water rises into the atmosphere as steam. The water eventually returns to the earth as precipitation—rain or snow—which finds its way into cracks in the rocks of the crust or becomes incorporated into certain types of rock-forming minerals. Over millions of years, as tectonic plates collide with other plates and are subducted, those molecules begin another slow rise to the earth's surface. Thus there is a geological water cycle akin to the hydrologic cycle by which moisture falls to earth from atmospheric rain clouds, evaporates, and returns to the atmosphere—except that the geological cycle proceeds at an infinitely slower pace.

The great heights attained by some volcanoes give evidence of the enormous pressures generated by rising magma. In the world's highest volcanoes, Llullaillaco and Cerro Ojos del Salado in the Andes of South America, magma has been pushed to altitudes, respectively, of 6,723 and 6,908 meters above sea level. Moreover, particles of magma in observed eruption columns sometimes reach heights of 30,000 meters or more. Most magma, however, never reaches the earth's surface. As much as 90 percent of the molten rock that enters the lithosphere remains at depth, where eventually it cools and solidifies. Even in cataclysmic eruptions, far more magma remains within the earth than erupts at the surface. Some 74,000 years ago, in what is now Indonesia, a volcano named Toba exploded with a colossal blast that hurled an estimated 3,000 cubic kilometers of pyroclastic material into the atmosphere. But that figure represents less than 10 percent of the volume of material—some 30,000 cubic kilometers—that is estimated to have been left behind in the magma chamber.

---

*Although for simplicity we use the term *water* here, in reality the "water" consists of dissociated atoms of hydrogen and oxygen, which combine to form water vapor (H_2O) only during an eruption.

Volcanic arcs

The angles at which tectonic plates are subducted generally range from 15 to 70 degrees, depending on the buoyancy of the subducting plates. Where subduction angles are shallow, the earth's curvature gives the plate boundary a shape like an arc of a circle, just as the rim of a dent in a rubber ball has a circular shape. Thus when magma generated along such a boundary rises through the overlying plate, it forms a curved row of volcanoes known as a volcanic arc. Volcanic arcs in the ocean form island arcs—for example, Japan and the Aleutian Islands of Alaska.

About 60 percent of the world's volcanoes on land—that is, those that have erupted on continents or, if in the sea, have risen above the surface—are in island arcs in the so-called Ring of Fire, a series of volcanic belts that virtually surround the Pacific Ocean above the plate boundaries shown in Figure 1-2. Another 20 percent of the active land volcanoes are in or near the Mediterranean Sea, where several small plates, or platelets, are colliding with one another. Because most of the earth's land is north of the equator, about two-thirds of the known volcanoes on land are in the Northern Hemisphere. There are more than 1,500 of these volcanoes, as catalogued in 1994 by Tom Simkin and Lee Siebert of the Smithsonian Institution in Washington.[5] More than 3,000 eruptions have been recorded during the past three centuries. Despite the large number of land volcanoes, they produce probably only 15 to 20 percent of the magma that reaches the earth's surface.

Oceanic ridges

For tectonic plates to collide in some places, they must diverge, or spread apart, in other places. Most spreading boundaries are within the earth's ocean basins, where they are marked by underwater ridges or mountain ranges many hundreds of kilometers wide. The Mid-Atlantic Ridge, for example, winds along the floor of the Atlantic Ocean, marking the boundary between the North American and Eurasian plates and between the South American and African plates. The axes of many oceanic ridges, notably the Mid-Atlantic Ridge, are elongated depressions called rift valleys, bounded on either side by faults. Most rift valleys are riddled with fissures, which provide pathways for enormous volumes of magma—probably 75 to 80 percent of the magma that rises to the earth's surface. The weight of overlying water prevents gases dissolved in the magma from escaping rapidly, so deep-sea eruptions are not explosive. The magma solidifies as part of the oceanic lithosphere, forming new crust.

Mantle plumes

Volcanism can also be manifested as plumes of hot mantle material produced by upwellings of heat originating deep in the earth. The magma that rises in plumes can surface through either fissures or volcanic conduits. The plumes can remain active for many millions of years, and they may be hundreds of kilometers in diameter. They create what geologists call *hot spots* on the earth's surface. Iceland lies over a hot spot within the rift zone between the Eurasian and North American plates. The islands of the Hawaiian archipelago, almost in the middle of the Pacific plate, were created as the plate slowly drifted northwestward above a stationary hot spot. Although mantle plumes have produced vast quantities of magma in the past, they are less productive today than other forms of volcanism.

Uncorking the champagne

The eruption of a volcano is often likened to the opening of a bottle of champagne. The dissolved gas (carbon dioxide) in champagne remains in solution as long as the bottle is tightly corked to keep the liquid under high pressure. But the moment the cork is removed and the pressure reduced, the gas separates from the liquid and expands suddenly (creating the "pop"), and champagne flows from the bottle (or erupts, if the bottle is opened carelessly) as a bubbly foam.

In a volcano, of course, the liquid is magma, which contains a variety of gases (mostly water vapor), all under great pressure. Whether a volcano erupts explosively or quietly is a function of the magma's viscosity. Just as highly viscous magma resists flowing, it also resists the separation of dissolved gases—until the magma reaches the earth's surface and the confining pressure is released. Then, as with a bottle of champagne, the gases expand suddenly and the volcano erupts convulsively, shredding the molten magma into myriad droplets that, upon cooling, become pyroclastic fragments.

If the magma has low viscosity and therefore flows readily, the gases are under much less pressure and separate easily from the molten rock. The result can be a relatively quiet eruption: the magma merely oozes from the earth. The viscosity of magma is directly related to its content of silica, or silicon dioxide, a common component of many minerals. The more silica, the higher the viscosity and the more sluggish the magma.

The volcanic explosivity index

To compare the magnitude of volcanic eruptions, geologists have developed a *volcanic explosivity index*, or VEI, similar in principle to the Richter scale for earthquake magnitudes. The index is based mainly on the volume of explosion products (Figure 1-4) and the height of the eruption cloud. Each succeeding category represents a tenfold increase in explosivity, or explosive power, over the next lower category.

Eruptions with VEIs of 0 or 1, like most of those in Hawaii, typically ooze lava with little or no violent activity. Explosive eruptions generally have VEIs of 2 to 5. But especially powerful eruptions like those of Bronze Age Thera in the eastern Mediterranean, Italy's Mount Vesuvius in 79 c.e., and

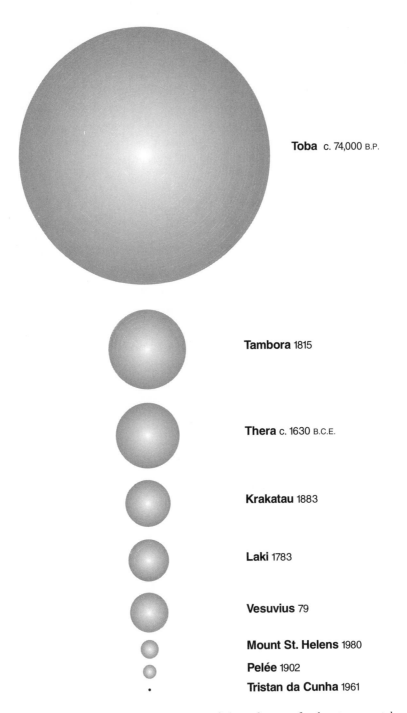

Toba c. 74,000 B.P.

Tambora 1815

Thera c. 1630 B.C.E.

Krakatau 1883

Laki 1783

Vesuvius 79

Mount St. Helens 1980

Pelée 1902

Tristan da Cunha 1961

FIGURE 1-4. Schematic comparison of the volumes of volcanic materials emitted during the eruptions discussed in this book.

TABLE 1-1. The major eruptions discussed in this book, in order of increasing VEI

VOLCANO	LOCATION	YEAR	INTENSITY	VEI
Tristan da Cunha	Tristan da Cunha	1961	Moderate	2
Surtsey	Iceland	1963	Moderate	3
Eldfell	Iceland	1973	Moderate	3
Kilauea	Hawaii	c. 1790	Large	4
Laki/Grimsvötn	Iceland	1783	Large	4
Pelée	Martinique	1902	Large	4
Mount St. Helens	United States	1980	Very large	5
Vesuvius	Italy	79	Huge	6
Thera	Greece	c. 1620 B.C.E.	Huge	6
Krakatau	Indonesia	1883	Huge	6
Tambora	Indonesia	1815	Colossal	7
Toba	Indonesia	c. 74,000 B.P.	Humongous	8

NOTE: B.C.E. means "before the commmon era"; B.P. means "before the present." The terms used to describe intensity are those employed by volcanologists. VEI stands for volcanic explosivity index.

Indonesia's Krakatau in 1883 probably had VEIs of 6. Tambora, also in Indonesia, erupted in 1815 with an estimated VEI of 7. And 74,000 years ago the colossal eruption of Toba, mentioned earlier, is thought to have had a VEI of 8 or higher (see Table 1-1).

Low-VEI eruptions are much more frequent than highly explosive eruptions. Volcanic events with VEIs between 0 and 3 may occur every few years somewhere on earth. In contrast, eruptions with VEIs greater than 6 occur at intervals of up to thousands of years.

The volume of material ejected by a single volcanic eruption can be prodigious. Many pictures have been published showing the huge cloud produced by the 1980 eruption of Mount St. Helens in the state of Washington. Impressive as it was, that eruption was but a burp compared with truly great eruptions of the past. In 1883 Krakatau emitted about eight times as much material as Mount St. Helens. The 1815 explosion of

Tambora produced at least thirty times as much material. And Toba is thought to have produced a thousand times as much material as Mount St. Helens, as illustrated in Figure 1-4.

Destructive power

Causes of damage from volcanic eruptions are not restricted to gas emissions, lava flows, pyroclastic flows, and ashfalls. Volcanoes are notoriously unstable mountains. Many rise thousands of meters above surrounding lowlands, with flanks so steep that minor earthquakes can cause massive landslides. Most large volcanoes are high enough that warm, moist air rising up their flanks forms clouds near the summit. Hence such volcanoes are subject to frequent rainstorms or, if high enough, snowstorms. Water-saturated mountain soils, as well as packed snow and glacial ice, are very likely to break loose and become landslides or avalanches. Both are commonly triggered by earthquakes associated with eruptions. Landslides coursing down the valleys of mountain streams often become transformed into mudflows, which can travel great distances at high speed, destroying everything in their path.

Moreover, many volcanic craters accumulate large volumes of water from rainfall or melted ice or snow. During the early stages of an eruption, as hot magma rises toward the surface, that water may become boiling hot. Eventually it may be forced from the crater by upwelling magma, causing hot, boiling mudflows that are terrifying, and fatal, for anyone caught in their path.

Major eruptions can change weather patterns, not only locally but also regionally and even globally. Eruptions with high VEIs pour enormous quantities of dust and sulfur dioxide gas into the atmosphere. The dark dust particles absorb sunlight. The sulfurous gas molecules react with atmospheric water vapor to form tiny droplets, or aerosols, of sulfuric acid. The light-colored aerosols reflect sunlight. Thus such eruptions reduce the amount of heat reaching the earth, and surface temperatures are lowered. Veils of volcanic dust and aerosols can remain in the atmosphere for years. Carried around the world by high-altitude winds, they can have serious long-lasting effects on global weather patterns. Because most land masses, hence most volcanoes on land, are north of the equator, the Northern Hemisphere is especially vulnerable to weather changes related to volcanism.

The destructive power of volcanoes is not limited to periods of eruption. Even extinct volcanoes are potentially dangerous. As they age, the mountains become more and more unstable. Eventually an entire flank, weakened by fractures, might collapse, causing a landslide of prodigious proportions. Or if the flank of a volcano should collapse into the sea, as has happened in the Hawaiian Islands, it would create a giant wave, or *tsunami*. Tsunamis can wreak havoc when they crash ashore on other islands or even on the shores of continents far across the ocean.

In this book we describe nine volcanic eruptions, which varied in the amount of destruction they caused and had effects on humankind, for good or ill, that ranged from local to global in scale. In each chapter, we briefly discuss the geological setting of the event and its immediate consequences. Then, as in our metaphor of the "vibrating string," we emphasize the most significant long-term aspects of each eruption—those aftereffects that have changed lives, societies, and cultures.

DATING OF VOLCANIC EVENTS

The accurate dating of volcanic events is crucially important in relating them to human endeavors. Most volcanic events that have occurred during historical time are reasonably well dated. The ages of earlier eruptions are less certain, and those events are dated by scientific methods that are still evolving. For example, eruptions that alter weather patterns can affect the growth of trees. Thus the width of annual growth rings can be an indication of aberrant weather, possibly caused by volcanic activity. The sulfuric-acid aerosols that form in the atmosphere after major eruptions eventually settle back to earth, and in glaciated regions they leave traces of acid in annual layers of ice. Thus cores taken from ice caps in Greenland and Antarctica have provided evidence of volcanism.

Although annual tree rings and acidic layers in ice cores can indicate time in terms of years, they cannot always be related to a specific volcanic eruption. But when molten lava cools and solidifies, its component minerals, some of which contain iron, often retain a magnetic orientation parallel to that of the earth's magnetic field at the time when the lava was molten. This phenomenon, called *paleomagnetism,* can be used to correlate the magnetic orientation of the solidified lava with different known directions of the earth's magnetic field in the past. Thus paleomagnetic studies can reveal the approximate time of a specific volcanic eruption.

Another widely used dating method is to measure the amount of radioactivity given off by isotopes of certain chemical elements. The most common of these radiometric methods is to analyze the carbon in an organic substance and determine the amount of carbon-14 relative to the amount of carbon-12, the most common isotope, in a given sample. Cosmic rays entering the earth's atmosphere react with atmospheric gases, and one of those reactions changes nitrogen to carbon-14 and hydrogen. Carbon-14 is radioactive, having a half-life of about 5,730 years. Both carbon-14 and carbon-12 react with oxygen in the atmosphere to form carbon dioxide, which eventually is taken up by living plants. When a tree, for example, dies or is cut down for firewood or lumber—or is killed in a volcanic eruption—it no longer takes in carbon dioxide, and the amount of carbon-14 it contains begins to decrease by radioactive decay. Therefore the ratio of carbon-14 to carbon-12 in a piece of the tree, in ashes from a fire, or in the timbers of a house provides an indication of how long ago the tree died. The lower the ratio, the older the eruption that killed the tree.

The carbon-14 dating method assumes that the rate at which that isotope forms in the atmosphere has remained constant for thousands of years. Although we know the rate has not in fact remained constant, this method is considered quite reliable as long as corrections are applied. Other, even more reliable methods make use of the relative proportions of different isotopes of argon, or of argon and potassium, in the minerals in volcanic rocks. So-called argon-argon and potassium-argon dating methods offer great precision and are especially useful for obtaining much older dates than can be obtained with carbon-14. Newer, less common dating methods are also available for dating volcanic rocks.

Layers of volcanic ash in sedimentary deposits can be dated geologically if we know the age of a deposit, as by identifying fossils of known age or knowing the rate at which overlying sediments were deposited. Moreover, we can use this method to determine the origin of the ash by comparing its chemistry with that of ash from a known volcano.

2 · The Hawaiian Islands and the Legacy of Pele the Fire Goddess

A people believing that Pele the Goddess
would wallow in fiery riot and revel
On Kilauea,
Dance in a fountain of flame with her devils,
or shake with her thunders and shatter her island,
Rolling her anger
Through blasted valley and flaring forest
in blood-red cataracts down to the sea!

Alfred, Lord Tennyson, "Kapiolani"

THE PEOPLE OF HAWAII have always lived with the perils of volcanism. The islands' earliest settlers, like early humans everywhere, attributed natural phenomena to gods. Myths about the gods taught each new generation about their volatile environment. Moreover, the ancient beliefs formed the basis of religious and ethical codes, including many taboos, which helped the Hawaiian people cope with their often dangerous surroundings.

The awesome results of volcanic eruptions, the Hawaiians believed, were the work of a fearsome deity named Pele, the goddess of fire. Many myths involving Pele were derived from observations of the profound natural changes brought about by volcanism and earthquakes. Pele, who could appear as either a beautiful maiden, or an ugly old woman, was irritable. When annoyed, she would stamp her foot, and the earth would shake. When enraged, she might hurl fiery boulders (volcanic bombs) at offending mortals, or she would generate streams of molten lava, which would spew from mountain-tops or from cracks in the flanks of a mountain and destroy everything in their path. Sometimes the lava would flow down to the sea and create new land. Pele's power was equaled only by that of her sister Namaka o Kahai, the goddess of the sea, who could quench Pele's fires and erode the land created by her cooled and hardened lava.

The people both feared and respected Pele, and they tried to assuage her capricious temper with offerings of food and small animals, along with garlands of flowers, called leis, and sacred ohelo berries. They would place these gifts at the edge of a lava flow or near the fire pit of an active volcano or would sometimes hurl them directly onto the red-hot molten rock. Later, tobacco and even bottles of brandy or gin were added to the votive offerings.

The ancestors of the Hawaiian people came from the islands of Polynesia sometime during the first millennium C.E., having sailed northward across the vastness of an unknown ocean in large outrigger canoes. In 1782 a chief named Kamehameha became ruler of the island of Hawaii, and in 1795, as King Kamehameha I, he united all the major islands under his rule. The Hawaiian monarchy lasted until 1893, when Queen Liliuo-kalani was forced to abdicate in a revolution supported by American businessmen. Hawaii was annexed to the United States in 1898 and became a state in 1959.

The Hawaiian Islands form a northwest-southeast-trending archipelago that extends some 2,400 kilometers across the north-central Pacific Ocean, from tiny Kure Atoll, or Ocean Island, to the island of Hawaii itself, the "Big Island" (Figure 2-1). The islands are the peaks of the earth's highest mountains, if we measure from their base on the ocean floor.

The Himalayas of Asia, of course, are the world's loftiest mountains. Their highest peak, Mount Everest, has an elevation of 8,848 meters above sea level. The highest of the Hawaiian peaks, the often snow-capped Mauna Kea (Hawaiian for "white mountain") on the island of Hawaii, is 4,205 meters above sea level—but the ocean floor at its base is about 5,000 meters deep.

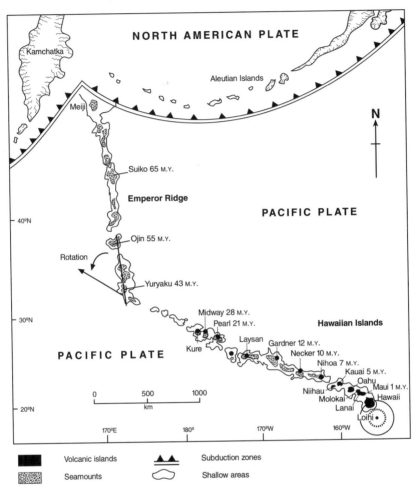

FIGURE 2-1. The Hawaiian Islands and the seamounts of the Emperor Ridge in the northern Pacific Ocean. Approximate ages are given in millions of years (M.Y.). The abrupt change in direction of this chain of volcanic features was created about 40 million years ago, when the Pacific plate changed direction as it passed over the Hawaiian hot spot.

Thus the true height of Mauna Kea is more than 9,000 meters. Neighboring Mauna Loa (long mountain) is only 35 meters lower than Mauna Kea.

The world's largest volcanic mountain in terms of sheer mass, Mauna Loa is an oval-shaped dome about 100 kilometers long and 50 kilometers wide at its base on the ocean floor. At its summit is a huge caldron-shaped depression, or caldera, known as Mokuaweoweo. It is some 6 kilometers long, 2.7 kilometers wide, and, in places, 180 meters deep.

The Hawaiian volcanoes have been built up very slowly, mainly by outpourings of lava, which, over millions of years, have covered large areas and created huge, dome-shaped mountains with gently sloping sides. Their shape is said to resemble a warrior's shield lying flat with its convex side up; hence they are known as shield volcanoes and are the proto-type for such volcanoes around the world.

During most eruptions in the Hawaiian Islands, lava pours from fissures in the flanks of the volcanoes. The eruptions typically last from a few days to several months, and several have gone on for years. Sometimes lava will flow around certain areas and leave them undamaged. These islandlike areas, which range in size from a few square meters to many square kilometers, are called *kipukas*. Many of them preserve plants from which seeds and spores spread to barren lava fields, where they start life anew.

The lava is relatively fluid, so volcanic gases ordinarily escape into the atmosphere with little of the ash, pumice, and cinders that are associated with eruptions that involve more viscous lava, from which gases escape explosively. Nevertheless, spectacular curtains of fiery lava, as well as volcanic bombs, may initially shoot into the air from several places along a fissure. Then the eruption settles down, and the emission of lava usually is restricted to one or more active vents. The molten rock flows down the mountainside in rivers that sometimes move as rapidly as 40 kilometers an hour.

The surface and bottom layers of a lava flow cool and harden as the lava moves down the mountainside. Sometimes solid tubes are formed, through which hot, liquid lava continues to flow. Some of these lava tubes are several meters in diameter. Mark Twain visited Hawaii in 1866 and, in his book *Roughing It,* described two such tubes:

Their floors are level, they are seven feet wide, and their roofs are gently arched. Their height is not uniform, however. We passed through one a hundred feet long. . . . It is a commodious tunnel, except that there are occasional places in it where one must stoop to pass under. The roof is . . . thickly studded with little lava-pointed icicles an inch long, which hardened as they dripped. . . . if one will stand up straight and walk any distance there, he can get his hair combed free of charge.[1]

Most lava solidifies into either of two forms of basalt known by the Hawaiian terms *aa* (pronounced "ah-ah") and *pahoehoe* ("pa-hoy-hoy"), which have been adopted by volcanologists worldwide. Aa has a rough, often jagged surface that is extremely difficult to walk upon, whereas the surface of pahoehoe is typically smooth and undulating, sometimes having a ropy texture. Where pahoehoe flows into the sea, it often forms masses of basalt that resemble piles of pillows.

Extensive lava flows, once hardened into basalt, tend to be highly permeable because of numerous intersecting fractures formed by shrinkage of the lava during cooling. The permeability of Hawaiian basalt is so great that most rainwater soaks directly into the ground. There is little runoff from these young flows. Well-defined streams do not have time to form, or if they do form, they are buried by new flows. On the older islands, however, where volcanism has ceased, many streams have eroded deep valleys. All the larger islands are blessed with large quantities of fresh groundwater, a priceless resource that is indispensable for Hawaii's agriculture and population.

Although some areas of the Hawaiian Islands are bleak and barren because of volcanic activity, or arid because high mountains shield them from rainfall brought by prevailing winds, the islands are mostly mantled with lush tropical vegetation. Mark Twain, after his visit in 1866, described the picturesque archipelago as "the loveliest fleet of islands that lies anchored in any ocean."[2]

The tectonic plate that underlies the Pacific Ocean is composed of dense, heavy basalt and is submerged everywhere except where island chains such as the Hawaiian archipelago, or Hawaiian Ridge, have risen above sea level. About 3,200 kilometers northwest of the island of Hawaii, the west-northwest-trending Hawaiian Ridge joins the north-northwest-trending Emperor Ridge, a chain of submerged volcanoes known as seamounts (see Figure 2-1). The abrupt change in direction—a rotation of about 35 degrees—was caused by a change in the motion of the Pacific plate. That change in turn resulted from changes in the orientation of centers of sea-floor spreading, where the Pacific plate is separating from plates that border it to the southeast. As shown in Figure 2-1, the Emperor Ridge can be traced to the juncture of the Kuril-Kamchatka and Aleutian trenches, along which the northwest-moving Pacific plate is being forced beneath the Eurasian and North American plates.

The origin of the submerged ridges and island chains in the Pacific Ocean remained an enigma until the 1960s, when a revolution in earth science led to general acceptance of the theory of plate tectonics and recognition of the mobility of the earth's crust. In 1963 a Canadian geophysicist, J. Tuzo Wilson, proposed that the Hawaiian Islands had been formed where the Pacific plate moved slowly west-north-westward over a hot spot—the surface manifestation of a plume of molten rock, or magma, that rose from deep in the earth's mantle. In 1972 W. Jason Morgan of Princeton University extended Wilson's concept to include the Emperor seamounts, which, he proposed, had been formed earlier above the same hot spot, but at a time when the Pacific plate was moving in a more northerly direction.

Wilson's concept of hot spots ignited a flurry of scientific activity that soon lent support to his idea. Oceanographers determined that the sea floor upon which the Hawaiian Islands have developed is 80 to 120 million years old and that the volcanic rocks exposed on the islands are much younger. The rocks increase in age northwestward, however, with increasing distance from the island of Hawaii and its currently very active volcano Kilauea. Haleakala, on Maui, last erupted about 1790

and is now considered dormant. Farther northwest, on Molokai, Oahu, and Kauai, the volcanoes are extinct. On Kauai the
volcanic rock is 5 to 6 million years old.

Starting at least 70 million years ago and continuing to
about 40 million years ago, the Pacific plate moved north-
northwest at an average rate of about 7 centimeters a year.
Since then the plate has moved west-northwest at a slightly
higher rate. Over time there has been an increase in the volume of erupted magma, as indicated by the large size of the
Big Island compared with other islands in the archipelago.

Volcanism in the islands appears to have occurred in dis-
crete pulses. Periods of volcanic activity have been separated
by periods of relative quiescence, and these pulses created a
chain of islands as the Pacific plate carried older islands
farther and farther to the northwest. The volcanoes on the
islands are arranged not in a single row, however, but in two
parallel rows about 40 kilometers apart, as shown in Figure 2-2. The two rows apparently formed on either side of
the stationary hot spot as the Pacific plate passed over it,
much as rows of ripples form downstream from either side of
a boulder in a river. In an article published in *The New Yorker*
in 1988, John McPhee elegantly describes the island-forming
process:

> Traveling upward in plumes of, say, two thousand miles,
> [magmatic heat] eventually encounters the thin surface
> plates, and may help explain why they move. In any case,
> when the heat reaches the underside of a plate it punches
> through, and, as the plate moves, punches through again,
> and soon again, like the needle of a sewing machine pene-
> trating moving cloth. . . . While the Pacific lithosphere slides
> overhead, the Hawaiian heat source stays where it is, mak-
> ing islands. There are five thousand miles of . . . islands,
> older and older to the northwest, reaching to the trench just
> east of Kamchatka. Almost all of them have long since had
> their brief time in the air, and have been returned by erosion
> and seafloor subsidence into the fathoms from which they
> arose.[3]

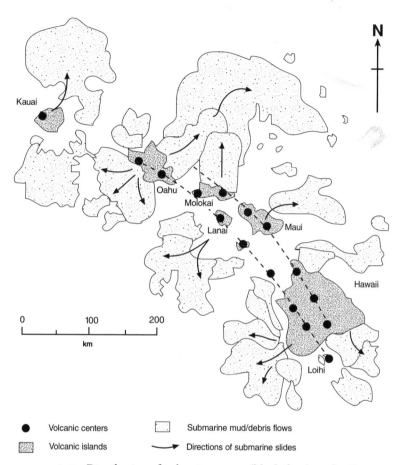

FIGURE 2-2. Distribution of volcanic centers (black dots) on the Hawaiian Islands (dark stippling). The volcanoes are arranged in two rows, as shown by dashed lines. Also shown (light stippling) is the extent of submarine landslides (mud and debris flows) resulting from gravitational collapse of large segments of volcanoes. Adapted from Moore et al., "Giant Hawaiian Underwater Landslides."

The model of a northwest drift of the Pacific plate over a stationary hot spot implies that once an island has left the hot spot, its volcanism should cease. That is not the case, however. Above the hot spot, the weight of volcanic material causes the lithosphere, the solid upper part of the mantle and the overlying crust, to subside, or settle downward. In turn, the subsiding lithosphere puts enormous pressure on the asthenosphere, the ductile part of the mantle that underlies the lithosphere. Under such pressure, the asthenosphere behaves like a plastic and is squeezed laterally. As the Pacific plate moves northwestward, then, the asthenosphere bulges upward ahead of the area of subsidence, forming a sort of "bow wave." As a result, the lithosphere ahead of the area of subsidence is uplifted.

The uplift causes crustal faulting, which creates pathways for renewed intrusions of magma. If the magma solidifies within a fault, it becomes a sheetlike body of rock called a dike. But occasionally the magma reaches the surface and erupts through a vent where, typically, a small cinder cone or spatter cone is formed. Thus on most older islands there is evidence of two phases of volcanism: a primary phase during which the massive shield volcanoes were formed and a minor, secondary phase represented mainly by swarms of dikes and surface vents.

Shield-building volcanism on Kauai, for instance, took place between 5 and 6 million years ago. But between 1 and 1.5 million years ago, volcanism returned with the development of swarms of north-northeast-trending dikes and vents. Shield-building volcanism on Oahu took place between 2.5 and 4 million years ago, but new upwellings of magma created northeast-trending swarms of dikes between 0.3 and 0.8 million years ago. This model implies that renewed volcanism can be expected on western Molokai in the near geological future, about a million years from now.

As a volcano grows older and moves away from the hot spot, it subsides at a rate that becomes faster than its rate of growth. Thus its elevation slowly decreases. Moreover, as soon as a new island rises above the sea, stream erosion, weathering, and wave action begin to wear it down—or, as it might be expressed in terms of Hawaiian mythology, no sooner does Pele create an island than Namaka o Kahai begins to destroy it.

Subsidence of the Hawaiian Islands has created a depression in the sea floor, just as a plank supported at each end bends downward when a heavy weight is placed upon it. The depression is evident on both sides of the archipelago. The still-growing Big Island of Hawaii is subsiding at a rate of 3 to 4 millimeters a year. Subsidence on Maui, farther to the northwest, is only about 2 millimeters a year. Oahu, still farther northwest, appears to have attained equilibrium; its load no longer depresses the mantle. In the near geological future, however, Oahu will start to sink again. The lithosphere beneath the island eventually will escape the buoyant effects of Hawaii's hot plume. The lithosphere will cool, become thicker (hence heavier), and slowly subside.

Northwest of Oahu, as the lithosphere moves away from the hot spot, it cools, becomes more dense, and gradually subsides, resulting in a progressive reduction of volcano height. The islands eventually submerge and become seamounts. Toward the northern end of the Emperor Ridge, the combination of subsidence and erosion has created a number of seamounts with flat, wave-eroded tops. Such features are known as *guyots* in honor of a pioneering Swiss-American geographer and geologist named Arnold Guyot.

Other old volcanoes, barely above sea level, have been reduced to rocky islets, and fringing reefs of coral have grown up around many of them. Charles Darwin was the first to point out that coral reefs typically continue to grow as a volcano subsides, creating the low, roughly circular groups of islands known as atolls, such as Midway.[4] The Emperor seamounts farther north have sunk too deep, and the water is too cold, to support coral growth.

The ridge that extends from the Hawaiian Islands to Midway contains eighty-two volcanoes with ages that increase northwestward to 28 million years. The youngest is Loihi, an active seamount about 30 kilometers off the south coast of Hawaii. It rises some 3,500 meters above the ocean floor on the submerged flank of Mauna Loa. Its top is still more than a thousand meters below sea level, however, so it will be a long time (probably hundreds of thousands of years) before it will breach the surface. Then it will form the next island in the chain or, more likely, merge with the island of Hawaii.

Ages for the submerged volcanoes of the Emperor Ridge have been difficult to determine. Volcanic rocks and sediments dredged from several of the seamounts have provided ages that increase north-northwestward from 40 million to 65 million years. Fossils with an age of about 70 million years have been found in sediments covering the Meiji seamount, the most northerly eruption center. Therefore the Hawaiian hot spot appears to have been creating volcanic islands for at least 70 million years. It may well have generated even older islands that have disappeared into the Kuril-Kamchatka Trench and are being melted back into the earth's mantle. If that is the case, many geologists believe the hot spot may have originated about 100 million years ago, roughly when many other hot spots are thought to have been created around the world—a time when the earth must have experienced a tectonic event of truly cataclysmic proportions.

———————

There are remarkable similarities between the origin of the Hawaiian Islands as explained by plate tectonics and the mythological explanation of how the fire goddess Pele came to live in Hawaii. The myths begin, again with geological verisimilitude, by recounting the birth of Pele. Her mother was Haumea, who personified the earth. Pele emerged from Haumea as molten lava. Her father was Ku-waha-ilo, the "man-eater," who represented the destructive forces of nature. Pele and her sister Namaka o Kahai were born on a mythical island somewhere in the South Pacific. Just as fire and water are incompatible, the sisters were always in conflict. To escape this sibling rivalry, Pele sailed away from their homeland in a great canoe provided by her brother Ka-moho-alii, the shark god. Namaka went off to a high peak on another island, where she could command all the seas.

Pele had a magic digging tool called Paoa. Wherever she landed, she struck Paoa into the earth and opened a volcanic crater where she could live. At first these small volcanoes were near the seashore, on the flanks of mountains rather than at their tops, and Pele's sister the sea goddess inevitably sent waves that doused the fires. In the Hawaiian Islands, many of them were formed by the explosive interaction of magma and seawater that seeped through fractures.

Eventually Pele made her way to Niihau, the northernmost of the main islands in the Hawaiian archipelago. But Namaka o Kahai would not allow her to remain there, so Pele moved eastward to neighboring Kauai, where she dug a deep fire pit. But again Namaka drove Pele from her home.

Pele next went to Oahu and dug a fire pit near present-day Honolulu, but her sister drowned the pit in salt water. Moving southeast along the shore, Pele used Paoa to dig a pit at today's Diamond Head. That landmark is now understood to be the eroded remains of an ancient, extinct volcano comprising many layers of compacted volcanic ash and fragments of a limestone reef that was penetrated by the volcano's upwelling magma. Its name comes from the glittering of calcite crystals in pieces of limestone as sunlight plays upon them. But alas! Pele dug too deep and struck water, which quenched her fires. What happened in fact is that groundwater or perhaps seawater penetrated into the magma chamber, flashed to steam, and created an explosive eruption.

So Pele next went to Molokai, where, on the north shore, she dug the crater known today as Kauhako. Lava emitted from that volcano created the isolated Kalaupapa Peninsula, site of a former leper colony, now a national historical park. But again Pele struck water. Namaka o Kahai forced her to move on, ever southeastward.

On the neighboring island of Maui, Pele at last climbed to the top of a mountain and succeeded in digging an enormous crater, today's Haleakala.* Namaka o Kahai saw smoke rising

*Today the site of a national park, the awesome depression was long thought to be a volcanic caldera. There is little doubt that it was originally volcanic (even today it contains lava flows and cinder cones), but in 1985, geologist Harold T. Stearns proposed that, as it exists today, the depression is an erosional feature caused by the joining of two valleys from opposite sides of the mountain. Stearns's view is now generally accepted.

from the mountain and came to do battle with her hated sister. In a furious fight she tore Pele to pieces and scattered her bones along the shore, where today there are irregular masses of lava called *Naiwi o Pele* (Pele's bones). Namaka o Kahai reveled in her apparent victory, but later, from her remote aerie, she looked back toward the island of Hawaii and saw clouds of red-hued smoke rising from a flaming crater atop the high mountain we call Mauna Loa. In those clouds she saw the spirit of Pele, and she knew then that she would never vanquish the goddess of fire.

It is said that when Pele arrived on Hawaii, she found another fire god, Ailaau, already in residence, but he fled before her greater power. His name, meaning "one who devours forests," alludes to the lava flows that frequently ate their way through the island's woodlands, setting them afire and burying mountainsides and fertile valleys. A few hundred years after such a catastrophe, however, the hardened lava would begin to break down from weathering, and its mineral nutrients would be added to the soil. New forests would grow, and crops would flourish in the valleys. Thus the Hawaiians, like people of many cultures in volcanic regions, recognized a duality in their fire gods. Their gods could be beneficent as well as destructive. Therefore the gods were both feared and respected, and even regarded with some affection.

The southeastward movement of the fire goddess in search of ever newer homes corresponds wonderfully to the northwestward movement of the Pacific plate over a hot spot that continually creates new volcanic islands. And the ravages of Pele's sister the sea goddess reflect a folk awareness of the erosional processes by which new islands become old and eventually, inevitably, disappear beneath the sea.

According to the mythology, Pele dwells today in Halemaumau, a fire pit within the caldera of Kilauea, located on the southeastern flank of Mauna Loa. Though it arises from Mauna Loa, Kilauea is a separate mountain. The two volcanoes have different conduits for upwelling magma, though occasionally magma from one volcano gets into the plumbing system of the other. Kilauea's summit is 1,247 meters above sea level, almost 3,050 meters lower than the top of Mauna Loa. Yet Kilauea is a huge mountain, rising some 6,100 meters from the ocean floor. Its oval-shaped caldera, almost 5 kilometers long and more than 3 kilometers wide, is bounded by sheer walls that are more than 120 meters high in places and made up of steplike fault blocks.

The caldera is famous for its fire pit, Halemaumau. At times, Halemaumau has contained a lake of molten lava. Forming a circular depression in the southwestern part of the caldera, the fire pit marks the top of Kilauea's conduit. Its dimensions vary, depending on volcanic activity and the occasional collapsing of its walls. In historical times, it has been more than 900 meters wide. Its bottom rises and falls according to the rising and falling of magma levels in the conduit. Sometimes lava boils up and overflows onto the floor of the caldera; at other times, when magma leaks into fissures in the mountainside, the lava lake disappears.

The lake presented an awesome spectacle of boiling, red-hot lava during much of the nineteenth century and well into the twentieth, making Kilauea unique among volcanoes. In 1924, however, a flank eruption—through the side of the mountain—drained Kilauea's lava lake, allowing water to enter the upper part of the conduit, where it flashed to steam and caused violent explosions. Today the floor of the fire pit is just a gray crust of hardened lava. Halemaumau's fires can be seen only during eruptions, when lava rises through fissures in the caldera floor or sometimes bursts into the air in fiery fountains. During these outbursts, droplets of molten lava may cool into globules of volcanic glass known as Pele's tears. Often, as they are wafted through the air, the droplets trail glassy threads, which break off upon cooling. Clusters of the spun-glass threads are said to be Pele's hair.

Mark Twain visited the caldera of Kilauea after dark one evening in 1866 and in his *Letters from Hawaii* wrote the following account:

The greater part of the vast floor of the desert under us was as black as ink, and apparently smooth and level; but over a mile square of it was ringed and streaked and striped with a thousand branching streams of liquid and gorgeously brilliant fire! . . . Imagine it—imagine a coal-black sky shivered into a tangled network of angry fire!

Here and there were gleaming holes twenty feet in diameter, broken in the dark crust, and in them the melted lava—the color a dazzling white just tinged with yellow—was boiling and surging furiously. . . . Occasionally the molten lava flowing under the superincumbent crust broke through . . . like a sudden flash of lightning, and then acre after acre of the cold lava parted into fragments, turned up edgewise like cakes of ice when a great river breaks up, plunged downward and were swallowed in the crimson caldron.[5]

Twain had something to say, too, about the sound and even the smell of lava: "the noise made by the bubbling lava is not great. . . . It makes . . . a rushing, a hissing, and a coughing or puffing sound. . . . The smell of sulfur is strong, but not unpleasant to a sinner."

Kilauea, the youngest of the Hawaiian volcanoes to have emerged above sea level, is among the most active volcanoes in the world. Its eruptions have been frequent, with periods of repose typically lasting from a few days to several decades. Many eruptions are preceded by inflation, or swelling, of part of the mountain as magma penetrates into fractures in the volcano's flanks. It is as if the volcano inhales. The swelling changes the slope, or tilt, of the mountain's flanks. Such tilting is imperceptible except to tiltmeters, sensitive instruments that measure minute changes in the vertical position of one point with respect to another. Upward passage of the magma is usually accompanied by small earthquakes. Tiltmeters that detect a mountain's inflation, and seismographs that detect pre-eruption earthquakes, are important tools in efforts to predict volcanic eruptions.

Although most eruptions of Kilauea are lava flows that ooze from fissures, explosive eruptions are not unknown. In

1916 Thomas Jaggar, director of the Hawaiian Volcano Observatory on Kilauea, described an eighteenth-century eruption that now is estimated to have had a volcanic explosivity index, or VEI, of 4—explosive enough to be destructive and dangerous. Wrote Jaggar, "even Kilauea is not guiltless of terrific and destructive explosive eruption. About 1790, thousands of tons of gravel and boulders and dust were strewn over Hawaii from Kilauea, covering hundreds of square miles, destroying the vegetation, and killing some of the people."[6] At the time of the 1790 eruption, the island of Hawaii was ruled by Kamehameha, the chief who, five years later, became king of all the Hawaii Islands. Kamehameha's right to rule the island was being challenged by another chief, named Keoua. From the town of Hilo, on the east coast, Keoua set out to cross the island with a small army and attack Kamehameha's forces near the opposite shore. Their path took them past Kilauea, and possibly as a precaution against the volcanic activity, Keoua divided his army into three groups.

The volcano erupted violently just after the first group, led by Keoua himself, had passed the caldera. The ground shook, and Kilauea ejected a dense cloud of gas, ash, and cinders. The hot debris killed some members of Keoua's party and injured others. The third group, the rear guard, though closest to the caldera, was not in the cloud's path, so members of that group escaped injury. Rushing forward, they came upon the second group and found them all dead, though there were few signs of physical injury. Probably they had been asphyxiated by a cloud of volcanic gas and steam that hugged the ground. Such phenomena, called *base surges,* often accompany eruptions caused by the penetration of groundwater into a volcano's conduit. The water reacts violently with magma and explodes into steam.

Kilauea continued to be active during most of the period from the early 1800s until the flank eruption of 1924, mentioned earlier. In 1868 there was an eruption that was remarkable because it consisted not of lava, but of mud. Known as "the great mud flow," it accompanied a severe earthquake and has been described as follows:

In the midst of the great earthquake we saw burst out of the top of [a precipice] . . . an immense river of . . . red earth . . . , which rushed down in headlong course and across the plain below, . . . swallowing up everything in its way—trees, houses, cattle, horses, men, in an instant as it were. It went three miles in not more than three minutes' time.[7]

Most of Kilauea's nineteenth- and twentieth-century activity, however, has involved lava flows from large fissures and from Halemaumau, the lava lake. In 1983 an eruption began from a fissure on the volcano's southeast flank. Known as the Pu'u O'o eruption, from the name of the vent where it began, it continues as this is being written eighteen years later. In 1986 its locus moved 3 kilometers eastward along the fissure zone. Lava production was virtually continuous from that site until 1992, when the locus shifted back to Pu'u O'o. Lava from the eruption has destroyed several communities on Kilauea's southeast flank.

Cape Kumukahi, the eastern extremity of the Big Island, has been created by repeated lava flows from Kilauea. In Hawaiian mythology, a story about the origin of the cape again illustrates the many correlations between geology and ancient beliefs about the fire goddess Pele. Cape Kumukahi is named after a mythical Hawaiian chief who committed the fatal error of refusing Pele when, appearing as an old woman, she asked to take part in some royal games. When the chief haughtily refused her request, a fountain of fire burst from the ground. Kumukahi ran toward the sea, but the vengeful Pele trapped him on the beach in a lava flow, which continued on into the sea and created the cape.

Another chief, Papalauahi, and several friends, suffered Pele's wrath when they beat her at the sport of *holua* riding. A *holua* is a kind of sled on which Hawaiian youths race down specially prepared grass-covered slopes. Pele saw to it that the racers were caught in a lava flow and turned into pillars of rock—in reality tree molds, which sometimes are created when a stream of lava flows through a forest and solidifies around large trees. The trees burn out, leaving the lava in the form of a hollow cylinder.

Holua riding figures in another myth, this one involving a chief named Kahawali, an expert *holua* racer. Again Pele appeared as an unprepossessing woman and asked to use Kahawali's sled. Curtly refusing, he sped off downslope. The angered Pele stamped on the ground, and an earthquake caused the hill to split open. Lava gushed forth and flowed down the race course, with the fiery goddess racing atop the molten rock on her own sled. Kahawali ran to the sea, leaped into a canoe, and furiously paddled away from shore. Pele's advance was stopped by the water, so she hurled rocks (volcanic bombs) after Kahawali, although, so the story goes, she failed to kill him. This myth provides not only a lesson in civility but also a graphic description of a flank eruption near the seashore.

The conduit through which magma rises into Kilauea has been traced by means of seismic data to a depth of about 35 kilometers. Down to that depth, the conduit appears to have a central core surrounded by numerous radiating fissures. Magma rises through both the core and the fissures during highly active periods, but through the core alone during periods of little activity.

Magma does not originate at a depth of only 35 kilometers. Melting begins much deeper in the mantle. Geochemical models suggest that the major melting region of the Hawaiian hot spot is about 80 kilometers deep. The volcanic rocks at the surface, however, carry traces of certain isotopes of helium that we know must have originated at depths greater than 300 kilometers. Therefore the process that ultimately leads to the formation of hot spots must begin deep inside the earth.

The present Hawaiian hot spot has a diameter of about 80 kilometers. The lighter molten material appears to be drawn to several conduits in the middle of that zone and may accumulate in large volumes below the volcanoes. As new batches of magma enter the conduits, upward pressure increases, and magma that arrived earlier surges toward the surface. There, in the crater or caldera of the volcano, lava may accumulate in large lakes, which often drain along fissures radiating from the volcano's summit. In contrast to calderas formed by explosive volcanoes such as Indonesia's Krakatau and Italy's Vesuvius, which are on the margins of converging tectonic plates, most Hawaiian calderas developed when the crater walls collapsed as the magma drained away.

Despite the shieldlike shape of Hawaiian volcanoes, with their gently sloping flanks, many of the islands are bordered by steep scarps, some of them hundreds of meters high. Some form spectacular cliffs on the islands themselves (most notably on Kauai and Molokai), and some are hidden beneath the sea around individual islands (Hawaii) or groups of islands (Oahu, Molokai, Lanai, and Maui). Known as *pali* in Hawaiian, these scarps represent eroded faults along which large slices of the volcanoes have slipped catastrophically toward the deep ocean floor—a process known as gravitational collapse.

The massive blocks that were detached along such faults caused submarine landslides (see Figure 2-2). For example about 105,000 years ago, as determined by the age of over-lying sediments, a large crustal mass was detached from an underwater scarp southwest of Lanai. The slide, about 40 kilometers wide at the top, became dislodged and traveled as far as 70 kilometers southwestward along the volcano's flank, where it came to rest on the ocean floor at a depth of about 4,500 meters.

That submarine landslide created a series of giant waves, or tsunamis, that must have reached the south coast of Lanai in about five minutes. The first wave crashed ashore and rose as high as 190 meters above sea level, as indicated by chunks of coral reefs and large pieces of ocean-floor basalt that broke off and were carried far inland. A second wave arrived one or two minutes later and reached an elevation of 375 meters on the island, carrying with it much debris from the first wave. A third wave was again about 190 meters high, and subsequent waves became gradually lower. As the waves retreated from Lanai's south coast, they stripped away all vegetation and soil up to the highest elevation they reached, leaving a rocky desert. Waves striking the north coast of the island stripped the ground to an elevation of about 100 meters, which shows that tsunamis can wrap around an island and do considerable damage on the side facing away from their source.

The events described above illustrate an often-overlooked aspect of volcanism—the fact that volcanoes, even when no longer active, can be dangerous. Many old volcanoes literally fall apart. The resulting landslides and the related tsunamis can have devastating consequences for residents of coastal areas.

Along the northeastern coast of Oahu, the seaward flank of the massif known today as the Koolau Range sheared off in a sudden cataclysmic landslide that strewed large blocks of volcanic rock several hundred kilometers offshore and left spectacular cliffs 500 to 700 meters high. Some individual blocks in that slide were the size of New York's Manhattan Island.

Similar slides have occurred off the Big Island along the western flank of Mauna Loa, where individual lobes have traveled as far as 100 kilometers beneath the sea, to depths of 4,800 meters. Estimates of debris thicknesses range from 50 to 200 meters, suggesting volumes of 200 to 600 cubic kilometers. The cumulative volume of the landslide terrain west of Hawaii is probably 1,500 to 2,000 cubic kilometers. To the east, the entire southern flank of Kilauea is slowly creeping southward along a major fault zone at a rate of several centimeters a year. Short periods of acceleration have caused earthquakes. In 1975 the mass lurched seaward by as much as 7 meters. The lower, submarine section of the slide displaced seawater, producing a tsunami that claimed two lives.

Earthquakes not only can result from landslides, they can also cause them. During the past century, there have been twenty-five earthquakes strong enough to dislodge large slabs of rock from the flanks of Hawaii's volcanoes. In April 1868 a quake in the southern part of the island of Hawaii triggered a landslide that buried a village and caused a tsunami.

The terrible power of slide-generated tsunamis is illustrated by a myth concerning a huge rock in the ocean off Kaena

Point, the northwestern tip of Oahu. The rock is called Pohaku
o Kauai, the Rock of Kauai, because it resembles the basalt of
that island, which is about 120 kilometers from Kaena Point. A
giant demigod named Hau pu, famed as a powerful warrior,
lived on Kauai and was awakened one night by the shouts of
men. He looked out over the sea and saw lights dancing on
the waves, far away toward Oahu. The men were fishermen
who had encircled a school of fish and were excitedly hauling
in their nets by the light of torches.

Assuming that the men were warriors from Oahu coming
to attack him, Hau pu tore an enormous boulder from the
ground and hurled it into the circle of lights. It caused great
wreckage and loss of life and raised huge waves, which swept
ashore on Oahu and created a long point of land. The peninsula was named Kaena Point after the chief who was said to
have led the fishing party.

In terms of Hawaiian mythology, we might say that
between the fire goddess Pele and her sister Namaka o Kahai,
the goddess of the sea, Namaka is the more destructive. Her
tsunamis, sometimes crashing ashore without warning, have
in the past caused terrible damage to the coastal regions of
Pele's realm. There is no reason to doubt that they will be any
less destructive in the future.

Because the Hawaiian Islands are located in the central
Pacific, tsunamis generated there travel outward in all directions. Thus, at times they devastate not only the islands themselves but also coastlines all around the Pacific Ocean.

Just as tsunamis originating in Hawaii can travel throughout the Pacific, the islands are often visited by tsunamis that
originate far away, triggered by slippage along major faults in
the zones of crustal convergence that surround much of the
Pacific Ocean. Deformation of the sea floor by faulting, usually
of the type that generates high-magnitude earthquakes, sets
up wave trains that inevitably reach the islands and frequently
cause great damage.

A tsunami warning system now alerts coastal dwellers
when a large wave is approaching the Hawaiian Islands from

elsewhere in the Pacific. People can be evacuated from its path, and preparations can be made to minimize wave damage. When a tsunami is generated in the islands themselves, however, there is no time for evacuation, and damage can be much more severe.

Unlike tsunamis, very few volcanic eruptions occur today without advance warning. Thus at various times, in various parts of the world, it has been possible to employ human technology in efforts to control lava flows once an eruption has begun. On the island of Hawaii, an attempt at lava control was made during an eruption of Kilauea in 1960. A lava flow threatened the village of Kapoho, on the mountain's eastern flank. Bulldozers pushed up an earthen barrier some 6 meters high and 5 kilometers long, but lava eventually overflowed it and destroyed the village.

During an eruption of Mauna Loa that began on November 21, 1935, great quantities of molten lava collected in the saddle between that mountain and Mauna Kea and then broke loose. A fiery river more than 1,240 meters wide moved slowly down the valley toward Hilo, threatening to overrun the city. Thomas Jaggar, director of the Hawaiian Volcano Observatory at the time, decided to try halting it with explosives. He contacted the U.S. Army Air Corps base on Oahu, and on December 27 two squadrons of bombers attacked the advancing flow. Some thirty-three hours later the lava stopped moving, but it soon started again. Finally, on January 2, the flow trickled to a halt about 20 kilometers from Hilo. Scientists later questioned the effectiveness of the bombing, arguing that the lava flow had stopped because the eruption had ended.

Mauna Loa erupted again in April 1942, during World War II, and again sent a lava flow toward Hilo. Again the Air Corps was called in to bomb the lava flow. The bombers succeeded in breaking up the hard crust at the edge of the flow, and lava flowed through the openings. The lava merely flowed along beside the old stream, however, and rejoined it a short distance downhill. The flow did stop before reaching Hilo, but again the results of the bombing were considered inconclusive.

Hilo was threatened by yet another eruption of Mauna Loa in 1984, but that time the state government ruled that no attempt would be made to divert the lava flow. On the basis of previous experience, officials felt that the effort would be futile. Moreover, the legal consequences of diverting a lava flow onto otherwise undamaged property were considered much too formidable. Fortunately the flow stopped short of the city.

The difficulties involved in using technology to control lava flows encourage many Hawaiians to appeal to Pele, as they have done for generations. Traditionally failure to treat Pele with great deference has been considered unlucky. Two hundred years ago that belief was dramatically illustrated by events related to the last eruption of Hualalai, on the island of Hawaii. In 1801 lava issuing from the base of the volcano destroyed a number of villages and plantations and filled in a bay on the west side of the island. The lava also threatened some private fish ponds belonging to King Kamehameha. The king tossed offerings of breadfruit and fishes onto the glowing lava, but Pele was not impressed by those meager offerings from so exalted a personage. The lava flowed on. Then the king brought a pig and cast it in; the flow continued. Finally Kamehameha cut off some of his own hair and threw it on the lava, symbolically offering part of himself. Pele accepted that act of obeisance, and the eruption ceased.

That Pele was not all-powerful, however, was demonstrated twenty-three years later by Kapiolani, wife of a chief on the island of Hawaii. American missionaries had converted Kapiolani to Christianity, and she was determined to show that Pele was not to be feared, that the fire goddess was merely a superstition. In December 1824, despite dire warnings from priests of Pele, Kapiolani and a determined group of supporters made the arduous journey to Kilauea from her home in western Hawaii. Members of the old faith tried to dissuade her, fully believing she would die a horrible death if she approached Halemaumau, the fire pit where Pele dwelt. "I should not die by *your* god," replied Kapiolani. "That fire was kindled by *my* god [italics added]."[8]

With her friends, she climbed to the caldera and descended to the very rim of Halemaumau. There she defied Pele by breaking a religious taboo. She picked some sacred ohelo berries, but instead of eating some and sharing the rest with the fire goddess by throwing them into the fire pit, as custom demanded, she ate them all. She further insulted Pele by hurling stones into the pit. Then she led her faithful companions in prayer, and they sang a Christian hymn. Afterward, to the amazement of onlookers, the group climbed out of the caldera unharmed.

In 1892 the British poet Alfred, Lord Tennyson, immortalized that courageous act in a poem he titled simply "Kapiolani":

When from the terrors of Nature
a people have fashioned and worship a Spirit of Evil,
Blest be the Voice of the Teacher who calls to them
"Set yourselves free!"

. . . Great and greater, and greatest of women,
island heroine, Kapiolani
Clomb the mountain, and flung the berries,
and dared the Goddess, and freed the people
Of Hawa-i-ee!

A people believing that Peelè the Goddess
would wallow in fiery riot and revel
On Kilauea,
Dance in a fountain of flame with her devils,
or shake with her thunders and shatter her island,
Rolling her anger
Thro' blasted valley and flaring forest
in blood-red cataracts down to the sea!

. . . .

. . . Kapiolani ascended her mountain,
Baffled her priesthood,
Broke the Taboo,

Dipt to the crater,
Called on the Power adored by the Christian,
and crying, "I dare her, let Peelè avenge herself"!
Into the flame-billow dash'd the berries,
and drove the demon from Hawa-i-ee.[9]

Though Kapiolani flung stones, not "the berries" as Tennyson wrote, his poem captures the boldness of Kapiolani's exploit in 1824. The old beliefs persisted, however. In 1881 an eruption of Mauna Loa threatened Hilo. Princess Ruth Keelikolani, a granddaughter of King Kamehameha I, approached the edge of the advancing lava front and, chanting ancient incantations, ritualistically offered Pele silk scarves, a variety of food dishes, and brandy. The lava flow did not stop immediately, but by the next day it had come to a halt short of the city. That act of Princess Ruth did much to keep the ancient beliefs alive.

And still today the Hawaiian people attempt to assuage the wrath of Pele by placing offerings for the fire goddess near lava flows, or even upon them. Food, bottles of brandy, items of clothing, and other gifts are often left near the rim of the caldera on Kilauea, where the spirit of Pele continues to dwell in the fire-pit called Halemaumau.

The myths of Pele and her malevolent sister Namaka o Kahai live on, just as volcanic eruptions, tsunamis, and their aftereffects will continue to threaten the inhabitants of what Mark Twain called "the loveliest fleet of islands that lies anchored in any ocean."

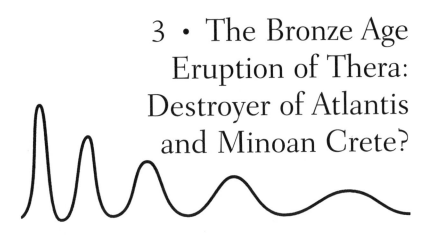

3 · The Bronze Age Eruption of Thera: Destroyer of Atlantis and Minoan Crete?

There was a time when the Mediterranean area was the hub of European, Egyptian, and Near Eastern civilizations, and during that time there was an eruption whose violence possibly has never been equaled in the memory of man anywhere on earth.

Dorothy B. Vitaliano, *Legends of the Earth*

A SMALL GROUP OF ISLANDS known collectively as Santorini lies in the eastern Mediterranean Sea about 110 kilometers north of Crete. The archipelago is roughly elliptical in outline and measures 13 by 18 kilometers. It comprises two large islands, Thera and Therasia, and three smaller islands named Aspronisi, Nea Kameni, and Palaea Kameni (see Figure 3-1). Santorini was named by early Venetians in honor of Saint Irene.

Unlike most other Greek islands, which dot the blue Aegean Sea with masses of white limestone and marble, Santorini's islands are dark and brooding. They are the remains of a great volcano that had its last major eruption more than 3,500 years ago, during the late Bronze Age. The eruption created a huge caldron-shaped depression, or caldera, which was superimposed upon other calderas that had formed during earlier eruptions. Thera, Therasia, and Aspronisi are parts

FIGURE 3-1. Bird's-eye view of Thera (after a drawing by Charles Lyell), showing the inundated caldera and an eruption of the Kameni islands in 1866. From Hull, *Volcanoes*, 79.

of the caldera's rim that remain above sea level. Between those islands the sea fills a deep, elliptical caldera measuring about 6 by 11 kilometers. Its greatest depth is more than 300 meters, and its sides—the caldera walls—rise in sheer, forbidding cliffs more than 400 meters high in places. Within the caldera the small islands of Palaea Kameni (old burnt island) and Nea Kameni (new burnt island), formed by post–Bronze Age volcanism, represent a new volcano in the making.

Thera, named for an early Greek conqueror named Theras, is the largest of the islands. We use this Greek name also for the ancient volcano, the Bronze Age eruption of which was certainly among the most devastating natural catastrophes in all of human history. Vast quantities of volcanic debris were hurled into the atmosphere. Much of it fell to earth on Thera itself, forming layers of ash and pumice up to 60 meters thick. The finer material was wafted by high-altitude winds as far as Crete, Asia Minor, and northeastern Africa. Earthquakes that preceded and possibly initiated the eruption are thought to have caused much damage to human habitations throughout the eastern Mediterranean.

Moreover, most geologists agree that when the caldera was formed, seawater poured into the abyss and created tsunamis that surged across the region in all directions. They probably devastated the coasts of many islands in the Aegean Sea, between Greece and Turkey, and overwhelmed coastal settlements on the mainland as well as on Crete.

———

The Santorini archipelago and the nearby island of Anáfi are the southernmost of the Cyclades islands, so called because a number of them form a rough circle. In the early Bronze Age, Santorini was known to the Greeks as Stronghyle, "the round one"—no doubt an allusion to its circular outline. Thera's Bronze Age communities evolved out of Neolithic settlements that thrived in fertile river valleys along the Aegean coast of present-day Turkey and also in parts of mainland Greece. We know the island was settled by Bronze Age people because remnants of stone walls and houses have been found beneath the ash and pumice on Thera and Therasia. Pottery found in the ruins shows that the Bronze Age culture was Minoan, from Crete. The large island of Crete developed a unique and powerful Bronze Age culture, which has come to be known as Minoan, after King Minos of Greek mythology.

The ancient Minoans had developed a written language known today as Linear A, samples of which have been unearthed on Crete—but no one has yet been able to decipher the script. And the Greeks had not yet evolved a written language when Thera erupted. Thus there is no historical record of one of the greatest geological disasters to befall humankind. Myths that have been handed down through the ages, however, almost certainly incorporate memories of the catastrophe.

The Greek myths of King Minos, Theseus and the Minotaur, Deucalion's flood, and even parts of the voyage of the Argonauts all may be rooted in ancestral memories of the eruption of Thera. The most famous of those ancient myths—the story of the "lost continent" of Atlantis—may reflect mixed memories of the violent destruction of Santorini (Thera) and the depopulation of nearby Crete, which the Greek historian Herodotus tells us occurred long before the Trojan War (c. 1200 B.C.E.). According to some authorities, the biblical stories about the plagues of Egypt also could have had their origin in events related to effects of the Bronze Age eruption. Each of these myths and accounts is discussed in more detail later in this chapter.

———————

Thera is one of a string of volcanoes that make up the Hellenic volcanic arc (Figure 3-2), which includes the Likades islands in the northern Gulf of Euboea in Central Greece, the islands of Aegina and Methana in the Saronic Gulf near Athens, the Aegean islands of Milos and Santorini (Thera), and Nyseros, an island near the coast of Turkey. Compared with volcanoes elsewhere, especially those in the Ring of Fire around the Pacific Ocean, the volcanoes of Greece are small and widely spaced. Their origin, however, is similar to that of their Pacific counterparts in that they are attributed to collision between plates of the earth's lithosphere—that is, the crust and the uppermost portion of the mantle.

The African plate is drifting eastward and simultaneously rotating counterclockwise. Thus northeastern Africa is swinging northward and colliding with the Eurasian plate. What is believed to be part of the African plate is being subducted beneath the Aegean Sea (see Figure 3-2). The tectonic setting of the Aegean area is complicated by motion of the Arabian platelet, which is moving northward and forcing the Anatolian platelet (Turkey) westward, toward Greece. Africa's northeasterly motion and the westward push of Anatolia are squeezing the lithosphere beneath the Aegean Sea and causing it to bulge upward in a sort of "geotumor," giving rise to the Cycladic islands.

This convergence of plates has produced one of the most tectonically active regions in the broad collision zone between

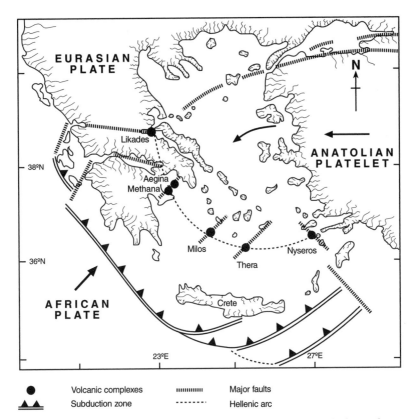

FIGURE 3-2. Tectonic setting of the Hellenic volcanic arc and Thera, showing the boundary where the African plate is being subducted beneath the Anatolian platelet, which includes the Aegean region.

Africa and Eurasia. As much as 5 percent of all the energy released globally by earthquakes comes from this relatively small part of the earth's lithosphere.

Subduction of the African plate has led to volcanism. At a depth of about 150 kilometers, the asthenosphere—the ductile, plastic portion of the mantle beneath the lithosphere—locally becomes molten. Blobs of magma then rise into the lower crust, intrude into faults caused by bulging and stretching of the Aegean lithosphere, and create the volcanoes of the Hellenic arc.

Thera is located in an area where one fault ends and another begins (see Figure 3-3). One, the Kameni fault, extends from

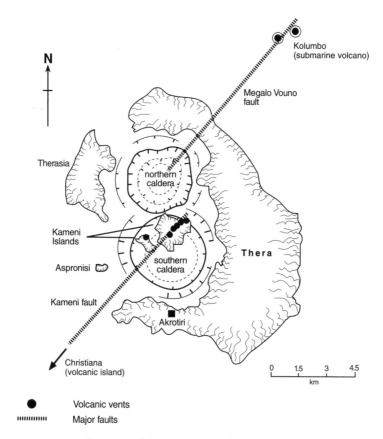

FIGURE 3-3. Diagram of the Santorini archipelago, showing modern Thera and the probable extent of calderas. The northern caldera is believed to be the site of the Bronze Age eruption. The Kameni islands first rose from the sea in the second century B.C.E. and continue to be volcanically active. Also shown are the Kameni and Megalo Vouno faults, which have provided pathways for ascending magma.

the volcanic island of Christiana, 18 kilometers southwest of Thera, to the Kameni islands in the center of Thera's caldera. The other, named the Megalo Vouno fault after a town on the north shore of Thera, extends northeastward from Thera to the submerged volcano of Kolumbo, about 10 kilometers from Thera, and from there it continues along the southeast coast of the island of Amorgos. The age of the most recent volcanic activity on Christiana is unknown because that island is privately owned and few geologists have had access to it. Kolumbo last erupted in 1650 C.E. Violent submarine explosions were accompanied by the emission of lava flows with sufficient volume to form a small island. Wave action, however, soon eroded the top of the new mountain to its present level below the surface of the sea.

————

As one enters the Santorini caldera by sea, the view is dominated by steep, towering cliffs, which largely surround the caldera and present a dramatic cross section of pale red and white volcanic ash and pumice alternating with layers of black lava. On the west side of the sickle-shaped island of Thera, the caldera walls reach a maximum elevation of more than 350 meters above sea level. This thick stratigraphic section is crowned by a layer of pink to white ash and pumice almost 60 meters deep in places, deposited during the Bronze Age eruption.

The rocks exposed in Thera's cliffs represent several hundred thousand years of intermittent volcanic activity. Although the age of the oldest volcanic deposits cannot be determined, since they are below sea level and are overlain by younger material, the island's volcanism probably commenced 1 to 2 million years ago.

The earliest major eruption for which dates are available occurred about 100,000 years ago above the Kameni fault. It left a deposit of pumice almost 30 meters thick and was followed by a second, more powerful blast that deposited almost 60 meters of pumice over much of central and southern Thera. Part of the volcano collapsed after those events and left a deep caldera in the south-central part of the island (see Figure 3-3). Another major eruption occurred in the same area about 80,000 years ago. Despite partial filling by volcanic products of later eruptions, the southern caldera is more than 250 meters deep.

The center of volcanic activity then shifted northward, and around 54,000 years ago an eruption above the Megalo Vouno fault deposited a layer of scoria, or cinders, almost 20 meters thick in places. Some 37,000 years ago (about 35,000 B.C.E.) another, much larger eruption produced a layer of scoria as much as 70 meters thick. As a result of those events, a new depression formed in the north-central part of what is today the Santorini archipelago. Around 16,000 B.C.E. that caldera was deepened by an eruption that left more than 40 meters of pyroclastic (fragmental) deposits on the islands. During the following fourteen centuries volcanic activity was restricted to the rising of new batches of magma into the Kameni fault and the growth of one or more volcanoes approximately where the Kameni islands are today. There were no more large eruptions until the Bronze Age.

The first settlers arrived on Thera sometime in the fourth or fifth millennium B.C.E. There, strategically located between Crete and the Greek mainland, they found an archipelago that virtually surrounded two large bays created by submerged calderas. Few other islands in the Cyclades had such well-protected harbors. Thera's red and black rocks and its steaming hot springs provided a strong contrast to the other Greek islands, with their white limestone cliffs. In these volcanic surroundings, the early settlers must have thought themselves close to the gods.

Then came the Bronze Age eruption that created the Santorini archipelago as we see it today. The eruption had four phases, as indicated by the succession of volcanic deposits. In the first explosion, ash and pumice were ejected high into the atmosphere and mostly fell back upon Thera and into the nearby sea. Those deposits have a maximum thickness of 6 meters on present-day Thera, and the thickness decreases southwestward to as little as 25 centimeters on the small island of Aspronisi. When the eruption column collapsed, it sent pyroclastic flows—clouds of superheated gases and fragmental material—bursting across the landscape. That first phase, along with earlier earthquakes, made Thera uninhabitable.

Data from archaeological sites on Crete suggest that the initial phase of volcanism was followed by a period, perhaps twenty years long, of relative quiet. Signs that a few buildings on Thera were repaired indicate that some people returned to the island during that hiatus, but they either left again or were killed by subsequent volcanic activity.

The second volcanic phase was characterized by steam eruptions. Seawater entered the bowels of the crater along faults, and when it contacted hot magma in the volcano's conduit, it erupted as powerful blasts of steam and ash. Those eruptions were followed by rapidly moving mudflows that formed deposits as much as 12 meters thick.

The third phase followed close upon the second and was the most destructive, as seawater probably entered the upper part of the magma chamber itself. The resulting blasts left ash, pumice, and large rock fragments in accumulations as much as 60 meters thick. The explosions must have been loud enough to have been heard throughout southern Europe, northern Africa, and the Middle East. Undoubtedly they blew such large volumes of volcanic dust and aerosols into the atmosphere that sunlight was dimmed over much of the eastern Mediterranean region for several days. These assumptions are based on comparison with the eruption of Krakatau, in the Sunda Strait between Java and Sumatra, in 1883. That explosion, with an estimated VEI (volcanic explosivity index) of 6, was heard as far as 4,600 kilometers away, and darkness in the region lasted for three days. Thera's VEI is assumed to have approached a value of 7, considered "colossal" by volcanologists.

During the fourth phase of the eruption the volcano poured out more pyroclastic flows, which left thick layers of ash, pumice, and rock fragments both on land and on the surrounding sea floor.

Estimates of the volume of the late Bronze Age eruption range from 30 to 60 cubic kilometers. This range represents the cumulative volume of volcanic matter blown from the caldera, including material from former eruptions and new material that rose from the magma chamber at depth. In view of the fact that earlier blasts may have helped excavate the center of Thera, the lower estimate appears most likely.

In any case, the Bronze Age blasts were tremendous. An enormous plume of volcanic debris, as much as 35 kilometers high, was blown toward the east and southeast. Up to 10 centimeters of ash fell on eastern Crete. A drill core from a lake in southwestern Turkey 320 kilometers northeast of Thera includes a layer of Thera's ash almost 13 centimeters thick. Since that layer was compacted by overlying sediments, its original thickness may have been double that. On the island of Rhodes, 210 kilometers east of Thera, a layer of the volcano's ash was found to range from 10 to almost 300 centimeters in thickness, a disparity no doubt resulting from the erosion of ash from higher elevations and its accumulation in depressions. The original thickness remains unknown but probably was at least 30 centimeters. Sediment cores collected in the Nile Delta include a layer containing fragments of volcanic glass from Thera, proving that the ash cloud reached Egypt.

The precise age of the Bronze Age eruption of Thera is in dispute in part because of uncertainties in the dating of ancient materials by the use of radiometry—that is, the measuring of radiant energy, in this case from a radioactive isotope of carbon known as carbon-14. Many questions have been raised about using carbon-14 because of insufficiently known variations in the amount of that isotope in the atmosphere over time. Moreover, fragments of charred wood, the most common source of carbon-14 samples on Thera, might have been contaminated by carbon-containing gases emitted from pyroclastic deposits. However, evidence presented at the Third International Congress on Thera and the Aegean World, held on Santorini in September 1989, indicates a statistical grouping of radiocarbon dates ranging from about 1690 to 1620 B.C.E. Data from ninety-four samples obtained from five different laboratories led to the conclusion that "the balance of probabilities" favors those dates in the seventeenth century B.C.E. for the eruption.[1]

A period of intense volcanic activity is known to have caused global cooling during that period. Volcanic eruptions eject minute droplets of sulfuric acid, which form aerosols in the atmosphere. Veils of aerosol droplets reflect sunlight, hence warmth. A peak in the sulfuric-acid content of annual layers in an ice core from Greenland indicates significant atmospheric pollution by volcanic aerosols around 1645 B.C.E. Frost damage and evidence of poor growth in tree rings from California bristlecone pines and from oak timbers found in Irish peat bogs indicate abnormal cooling between 1620 and 1630 B.C.E. Precise tree-ring records from ancient wood found in Turkey and Sweden provide dates, respectively, of 1628 and 1637 B.C.E.

Moreover, in China at about the same time, during the reign of the emperor Chieh, there were yellow fogs, most likely from sulfuric-acid aerosols, which dimmed the sunlight, according to ancient records. The seasons were abnormally cold, and there was frost in July. Abnormally high rainfall caused floods, which were followed by droughts and famine. All these phenomena correlate well with the time of Thera's eruption, which must have sent clouds of ash and aerosols over much of Asia.*

Many archaeologists have assumed that the demise of Minoan Crete, which they attribute to Thera's eruption, occurred two centuries later. Their evidence rests on linking the styles of ancient pottery found on Crete and in Egypt with historical dates from Egyptian hieroglyphic writings. Thus archaeologists have correlated late Minoan pottery with the reign of Thutmose III, the fifth pharaoh of the Eighteenth Dynasty, who is believed to have occupied his throne sometime between 1504 and 1450 B.C.E. The archaeologists' assumption depends on the accuracy and correct interpretation of ancient Egyptian texts. Rather than relying on such imprecise information, we should accept the globally correlatable scientific data. Further work on high-precision carbon-14 dating and more precise tree-ring information will, in time, undoubtedly provide a more accurate date for this important event.

---

*A disadvantage of the ice-core and tree-ring dates is that they cannot be directly correlated with Thera's eruption; that is, they could be related to some other volcanic event elsewhere.

Volcanism at Thera did not cease after the Bronze Age eruption. The islands of Palaea Kameni and Nea Kameni were formed long afterward inside Thera's caldera (see Figure 3-1). Palaea Kameni appeared above sea level in 197 B.C.E. There was more activity in 19 C.E., and another island emerged from the caldera's depths in 726 C.E. Still another island, called Micro Kameni (small burnt island) arose in 1570. Between 1707 and 1711 it merged with the second island to form Nea Kameni. New land emerged, sank, and re-emerged during eruptive periods in the late nineteenth and early twentieth centuries. The locations of the eruption centers indicate re-activation of the Kameni fault, allowing younger magma to rise through that zone of weakness. Earthquakes rocked the archipelago during and after those minor eruptions. Thousands of people left Thera after an especially severe quake in 1956, when scores were killed and many buildings were damaged. Much greater turmoil must have followed the Bronze Age catastrophe.

––––––––

Ships cannot anchor within Thera's caldera today because no ship carries enough cable to reach the bottom. Instead, when ships call at the island's main town, Phira, they tie up to mooring buoys anchored by enormously long chains. Cargo and passengers are ferried ashore in small boats. Until recently, to reach Phira, with its whitewashed buildings perched atop the caldera wall more than a thousand feet above the sea, one had to climb, by foot or on mule-back, more than 500 steps in a ramp that zigzags up the cliff. Today, however, cable cars whisk people to the top in minutes.

The view westward from Phira, looking out over the blue water in the caldera toward Nea Kameni and Therasia, is magnificent. Eastward, one looks down the gently sloping side of the old volcano, a mountain that one has literally climbed from the inside. Between the eastern shore and the rim of the caldera the landscape is a pleasant one of green vineyards and fields of tomato plants, barley, and beans, as well as scattered villages and a few roads. The volcanic soil is fertile, and crops grow well despite a dry climate. Water from winter rains is stored in cisterns, and in summer fresh water is brought to the island weekly by ship.

Thera is a favorite stop for cruise ships in the eastern Mediterranean. In addition to tourism, the island has one industry—the quarrying of pumice and volcanic ash. The largest quarry is just south of Phira, at the top of the cliff, where deposits from the Bronze Age eruption are about 40 meters thick. The powdery material is loaded into ships and taken to mainland Greece. There, in Athens, it is used in making a high-quality hydraulic cement, which sets under water. It was ash and pumice from Thera that went into the construction of the locks in the Suez Canal, which opened in 1869.

The basal layer of pumice in the Phira quarry rests on brown, loamy soil—the surface of the ground during the Bronze Age more than 3,500 years ago, when the volcano exploded and buried the island. Outlines of walls can be seen where the pumice has been removed in some places, and fragments of Minoan pottery have been found imbedded in the soil. The pottery fragments support an assertion by the historian Herodotus that King Minos of Crete colonized the Cycladic islands. Thucydides, another Greek historian, considered Minos' colonization of the Cyclades to be the beginning of Greek history.

Thera apparently was quite well populated at that time. Several Bronze Age sites have been uncovered in the ash and pumice quarries. Most have been interpreted as small towns or farms—country houses with walled-in animal pens. The first archaeological excavations were carried out on Thera and Therasia in the late 1800s. They revealed that there had been a prosperous Minoan colony on the islands, but little further archaeological work was done until 1967.

It was in that year that a Greek archaeologist named Spyridon Marinatos, excavating near the town of Akrotiri in southern Thera, discovered Bronze Age ruins of what has proved to be a city of considerable size—perhaps larger than Pompeii in

Italy. Like the ruins of Pompeii, the ruins of ancient Akrotiri consist of paved streets and the floors and walls of houses, some of them four stories high. Some of the walls, as in Pompeii, boast well-preserved frescoes and paintings. Many artifacts, largely Minoan in style, indicate close commercial contact with Minoan Crete.

———————

The Cretan Bronze Age began sometime around 3400 B.C.E., when artisans learned to mix copper with tin to make bronze implements and weapons. From Crete's then-extensive forests, Cretans obtained wood for the construction of ships. Excellent sailors, they traveled and traded throughout the eastern Mediterranean, soon becoming the dominant economic power in the region and establishing a hegemony over the Cyclades and most other islands as well. Moreover, they traded extensively with Egypt and Greece, even establishing colonies on the Greek mainland. Minoan Crete, in fact, though apparently not a politically aggressive nation, was powerful enough to subjugate Athens and exact tribute from the Greeks for many years. The ending of that period of Greek humiliation is attributed to Theseus, the mythological hero of Athens, and is discussed later in this chapter.

By about 2000 B.C.E., Cretan kings were living in elaborate palaces, mostly on the eastern half of the island. The greatest of the palaces was at Knossos, near present-day Iráklion. The ruins of the palace were discovered in 1895 by British archaeologist Arthur Evans. Because of similarities between the ruins and descriptions in Greek myths, Evans was able to identify the palace as the home of the mythical King Minos, after whom he named the Minoan civilization.

The Minoan culture ended sometime during the late Bronze Age. Palaces were destroyed throughout eastern and central Crete, agriculture came to a halt, and commerce ceased. Crete became depopulated, as Herodotus reported. People migrated westward from the eastern population centers, and many are believed to have emigrated to Greece or northern Africa.

There are no written records to tell us what happened, but the diaspora put an end to Minoan domination of the eastern Mediterranean world.

It was Spyridon Marinatos, the Greek archaeologist, who first proposed that the demise of Minoan civilization was related to the Bronze Age eruption of Thera. He set forth his theory in a now-classic article published in 1939 in the British periodical *Antiquity.* Marinatos suggested that earthquakes associated with the eruption had damaged Minoan sites in the interior of Crete and that tsunamis, caused by the collapse of the caldera, had devastated coastal locations, not only in Crete but also throughout the eastern Mediterranean. In support of his theory, he cited similarities to the destruction caused by the 1883 eruption of Krakatau in Indonesia.

Earthquakes related to volcanic eruptions generally have relatively low magnitudes, so their effects are not widespread. To cause destruction on Crete, Marinatos's quakes would have to have been of tectonic origin, caused not by volcanism but by the release of strain along major faults. The earlier quakes—those that preceded the Bronze Age eruption—did affect all of Crete. Damage apparently was concentrated in the eastern part of the island, where archaeologists have uncovered the ruins of many palaces and lesser buildings. Only the palace at Knossos survived, though it was badly damaged.

Many authorities have agreed with Marinatos, however, that the abrupt collapse of Thera's Bronze Age caldera did cause tsunamis. They could have been caused, too, by voluminous pyroclastic flows entering the sea from Thera. In either case, the waves would have traveled rapidly to nearby islands and continental shores. They would have been high enough to cause great destruction along the shores of Crete and Greece and even in the low-lying delta of the Nile in Egypt, where they would have penetrated far inland.

Recent studies have shown that some of the Bronze Age tsunami waves that reached Crete were more than 9 meters high.[2] The coasts of present-day Turkey and the Greek mainland were battered by waves almost as high. Such waves would have destroyed harbor facilities and any vessels in the harbors, just as tsunamis created by the collapse of Krakatau in 1883 destroyed harbors along the coasts of western Java and eastern Sumatra. For the Minoan empire, whose power resided in its commercial fleet, such damage would have been catastrophic. Even vessels at sea, which would not have been damaged as the waves passed beneath them, might well have been battered by strong winds and become stranded in huge fields of floating pumice. Seagoing vessels in Minoan times were oared galleys, and rowing would have been extremely difficult in such pumice fields.

In addition to tsunamis, eastern Crete was subjected to ashfall from Thera's eruption. On the basis of the thicknesses of ash in drill cores collected from the sea bottom off Crete, it is believed that a volcanic cloud drifted to the east and southeast and deposited at least 10 centimeters of ash on eastern Crete, probably less in the central part of the island. Vegetation would have been seriously damaged, and crops probably ruined, because rain would have washed the loose ash from higher elevations onto lowland farms, clogging irrigation systems and acidifying the soil. The resulting disruption of Cretan agriculture for several seasons would have led to famine and disease. Thus the ashfall was almost certainly the main cause of the depopulation of eastern Crete.

Still another factor—fire—was involved in the destruction of Cretan buildings. Blackened stones and layers of ashes leave no doubt that fire swept through virtually all the palaces, but the cause of the fires remains unclear. An early assumption that they were started by Thera's volcanism has been discounted because ash from the eruption, after being wafted 110 kilometers to Crete, would have cooled below any required kindling temperature. Yet paleomagnetic evidence from pottery and baked clay in walls and floors, fired during the destruction, shows that the ashfall and the fires were virtually contemporaneous.

In the past it has been thought that the fires started when earthquakes accompanying the eruption overturned olive-oil lamps, which would have been burning because of darkness caused by the ash cloud. Eventually the flames would have reached storerooms where supplies of oil were kept in large earthenware jars, which probably would have been cracked or overturned by the quakes, resulting in a general conflagration. It is now believed, however, that the larger earthquakes preceded the eruption, so there should be another explanation.

Perhaps there is. The eruption of Krakatau in 1883 created atmospheric percussion waves, or shock waves, that broke windows and cracked walls as far as 150 kilometers away. The distance from Thera to Crete is considerably less, and the Bronze Age eruption was more powerful. It is therefore quite possible that atmospheric shock waves, moving uninterrupted over the sea, overturned oil lamps and started fires when they slammed into Cretan buildings during the evening.

The destruction of Crete's palaces, the centers of power on the island, surely must have led to instability, social chaos, even anarchy. Many people fled the island. In Egypt's Nile Delta region between 1650 and 1520 B.C.E., there was a royal house of minor kings referred to as the Fourteenth Dynasty. They may well have been Minoan nobility or royalty who had fled from Crete.

Before the Bronze Age catastrophe, the culture of Crete was the nexus of Mediterranean civilization. To quote Homer, in the *Odyssey*, "There is a land called Crete, in the midst of the wine-dark sea, a fair, rich land, begirt with water, and therein are many men past counting, and ninety cities." In his acclaimed book *The Life of Greece*, Will Durant wrote,

> When Homer sang these lines, perhaps in the ninth century before our era, Greece had almost forgotten, though the poet had not, that the island whose wealth seemed to him even then so great had once been wealthier still, that it had held sway with a powerful fleet over most of the Aegean and part of mainland Greece; and that it had developed, a

thousand years before the siege of Troy, one of the most
artistic civilizations in history.

The rediscovery of that lost civilization is one of the major
achievements of modern archaeology. Here was an island
twenty times larger than the largest of the Cyclades, pleasant in climate, varied in the products of its fields and once
richly wooded hills, and strategically placed, for trade or war,
between Phoenicia and Italy, between Egypt and Greece.
Aristotle had pointed out how excellent this situation was,
and how "it had enabled Minos to acquire the empire of the
Aegean." But the story of Minos . . . was rejected as legend
by modern scholars.[3]

Then, in 1895, Arthur Evans began his archaeological
excavations at Knossos. Within the ruins he found clay tablets,
hardened and preserved by the fires that had destroyed the
palace. Writings on the tablets combined hieroglyphic pic-
tographs and an ancient, indecipherable script to which Evans
gave the name Linear A. Evans also classified shards of pottery and other relics from Knossos and compared them with
similar objects from ancient Mesopotamia and Egypt. On
the basis of accepted chronologies for the Mesopotamian and
Egyptian materials, he was able to divide the post-Neolithic
culture of Crete into Early, Middle, and Late Minoan periods.
Evans's work at Knossos led archaeologists to reassess the
historical value of the old myths.

The palace at Knossos, a maze of many rooms and intricate
corridors, very likely gave rise to the myths of the labyrinth,
the Minotaur, and Theseus, which are discussed later in this
chapter. Parts of the huge edifice rose to three or four stories.
Knossos was the grandest of the ancient Cretan palaces, but it
was only one of many. Their ruins, along with the remains of
ordinary houses, have been unearthed throughout eastern and
central Crete. Many of the houses had two or three stories, wide
windows, courtyards, and even toilets that could be flushed
into sewer systems.

Statuary, multihued pottery, and wall frescoes have yielded
a great deal of information about Minoan life. The frescoes

depict bucolic landscapes and both men and women with various animals, most notably bulls, which were considered sacred. Objects found in the ruins testify to exquisite craftsmanship not only in pottery but also in the fabricating of bronze basins, pitchers, and even daggers and swords, some inlaid with gold, silver, ivory, and gemstones.

There is archaeological evidence that the palaces of ancient Crete were damaged several times by earthquakes. Most of the palaces were rebuilt each time. In the early Late Minoan period, however, presumably at the time of the severe pre-eruption earthquakes, the destruction was so complete that no palaces were rebuilt—except Knossos, which seems to have suffered less damage. Even so, there were major cultural and social changes at Knossos. Pottery designs became less refined. And Linear A script gave way to a new system of writing. It was the same syllabic script as Linear A but was used for a different language, much as letters of our alphabet are used in several languages. Evans named it Linear B. In Crete it is known only at Knossos, but it has been found in ruins of a later date on the Greek mainland, in Mycenae, an ancient city in Peloponnesus. Just as Linear A has remained undeciphered, Linear B remained a mystery for over half a century. Finally, in 1952, a British scholar named Michael Ventris discovered that Linear B was an archaic form of Greek.

The implications of Ventris's discovery were of singular importance. The appearance of Linear B meant that not Minoans but Mycenaean Greeks had been in charge at Knossos at that time. With the end of Minoan dominance in the Aegean, Mycenaeans from Greece apparently had been able to invade Crete and occupy the palace at Knossos. Although many Minoans migrated into western Crete and even emigrated to Greece or northern Africa to escape the devastation wrought by earthquakes and the eruption of Thera, some must have remained at Knossos. Palace scribes who had been employed by Minoan kings to keep records of stores of grain, oil, and other essential materials at Knossos apparently modified their own language to accommodate the Greek words of their new rulers, who had no writing of their own. The Minoans thereby created Linear B, which Mycenaeans and emigrating Minoans took to Greece, giving that country its first written language.

Mycenaean culture in Greece benefited not only from the script imported from Crete but also from the creative talents of Cretan émigrés. In all probability it was the stimulating influence of the Minoans that led to a subsequent expansion of Mycenaean culture throughout the eastern Mediterranean.

———

There can be little doubt that a number of Greek myths have their roots in the Bronze Age eruption of Thera. Athens at that time was a Mycenaean city, and in Greek mythology it was Theseus, an Athenian, who slew the Minotaur on Crete and freed Athens (hence Mycenae) from domination by Minoan Crete—surely an allusion to the decline of Minoan culture and the concomitant rise of Athens and Mycenae.

According to the myth, the sea god Poseidon gave a magnificent bull to King Minos of Crete. The king's wife, Pasiphae, lusted after the bull. With the connivance of Daedalus, a Greek craftsman in Minos' service, Pasiphae seduced the animal. The result of their union was the Minotaur, a fierce creature that was half man and half bull, which Minos imprisoned in a maze, or labyrinth (the palace at Knossos), constructed at his command by Daedalus.

Minos imposed his authority on other nations, including Athens (the Minoan domination of the eastern Mediterranean), but his son Androgeus was killed by the Athenians. In revenge, Minos made Athens pay a yearly tribute of seven young men and seven maidens, who were sent to Crete and imprisoned in the labyrinth as prey for the Minotaur. One year Theseus, son of the Athenian King Aegeus (after whom the Aegean Sea is named), volunteered to go to Crete as one of the sacrificial youths. There he killed the Minotaur, freed the surviving Athenian hostages, and led them out of the labyrinth, ending the political and economic domination of Athens by

Crete. Could this myth refer to the demise of Minoan civilization as a result of Thera's eruption?

Mythical traditions of many cultures, from the Gilgamesh epic of ancient Babylon to the story of Noah in the Bible, refer to floods imposed by gods as punishment for human wickedness. The Greeks have the mythology of Deucalion's flood, which presumably occurred about the time of Thera's eruption and may reflect ancestral memories of the inundation of coastal embayments by tsunamis or heavy rainfall related to atmospheric pollution caused by the eruption. In the myth, Zeus decided to punish the wickedness of the time by destroying the world with a flood. The god Prometheus learned of Zeus' plan and warned Deucalion, who was Prometheus' son. Like Noah, Deucalion built an ark. When the waters rose, Deucalion and his wife Pyrrha took to the ark and survived. The Greeks, or Hellenes, were said to be descended from their son Hellen.

The myth of Jason and the Argonauts may well reflect Bronze Age events in early Greek history. After the Argonauts secured the golden fleece from Colchis, at the eastern extremity of the Black Sea, and were returning to Greece, they sailed through Minoan waters in the vicinity of Crete and Thera. Headed for Crete, they were prevented from landing by that island's guardian, a giant named Talos, made of bronze. He stood on a mountaintop and threw rocks at the seafarers. Hephaestus, the god of fire and metalworking, had fabricated Talos for Minos to keep intruders away from his kingdom. Talos would make himself red hot and embrace the intruders, killing them. Being made of bronze, he was invulnerable except for a place near one ankle, which was protected only by a thin membrane. The Argonauts fled from Talos' barrage of rocks, but Medea, a sorceress from Colchis who had married Jason and accompanied the Argonauts, cast a spell upon the giant to dim his vision. As Talos was preparing to hurl another boulder he stumbled and injured his unprotected ankle, and his vital fluids leaked away. He collapsed, fell from his mountaintop, and died.

It is not unlikely that Talos is a mythical representation of the island of Thera, which can be said to guard the northern approaches to Crete. Talos, in fact, was also known as Circinus (the circle), an apt description of the presumed shape of Thera in antiquity. The giant's invulnerable bronze physique could represent the volcano itself, and the rocks he threw might well be volcanic bombs. His red-hot embrace could represent a lava flow, and the weak spot at his ankle, a volcanic vent. His collapse and death as his vital fluids (lava) oozed out of him would be analogous to an eruption.

The Argonauts, after fleeing from Talos, sailed northward and were enveloped in a mysterious darkness. They escaped by appealing to Apollo, who guided them to the nearby island of Anáfi. The mythical darkness may refer to the dense gloom that must have been cast over the eastern Mediterranean by Thera's ash cloud.

Talos is supposed to have had a son named Leukos (the white one), who, according to the myth, drove away the king of Crete, destroyed many towns on the island, and killed the king's daughter, whose name was Kleisithera (key of Thera). The name Leukos probably refers to the white volcanic ash that covered eastern Crete and led to the diaspora of the Minoans.

Stories in the Bible, too, according to some authorities, may be rooted in the eruption of Thera. Refugees from Crete— called Caphtor in the Bible—are known to have settled in northern Africa, including Egypt and what is now Tunisia, and along the coast of Palestine, where they became known as Philistines. The Book of Amos (9:5–7) makes reference to those migrations and appears to link them to volcanism and floods, which may have been caused by tsunamis. The following quotation hints at such natural disasters, which triggered migrations of large ethnic groups on a regional scale:

> And the Lord God of hosts is he that toucheth the land, and
> it shall melt, and all that dwell therein shall mourn: and it
> shall rise up wholly like a flood; and shall be drowned. . . .

It is he . . . that calleth for the waters of the sea, and poureth them out upon the face of the earth. . . . Have not I brought up Israel out of the land of Egypt? And the Philistines from Caphtor . . . ?

Some authorities have suggested that the biblical plagues of Egypt, which preceded the Jewish Exodus, might have been related to the Bronze Age eruption of Thera. Biblical scholars, however, generally place the Exodus in the fifteenth century B.C.E., whereas the eruption of Thera, as we have seen, most likely occurred much earlier, in the seventeenth century. Both centuries contained periods of political instability in Egypt that could have led to the release or escape of Hebrew slaves.

Among the plagues, as described in the Book of Exodus, were waters turning red, thunder and hail, and darkness. All these phenomena could well have been of volcanic origin, whether from Thera or from a later volcanic eruption. Another possible source of the plague stories might have been earlier Egyptian records of similar events. Just as there are similarities between the biblical account of Noah's flood and the earlier Greek myth of Deucalion's flood, there are similarities between the biblical plagues and accounts of events that befell Egypt much earlier. As related on a papyrus scroll now in the National Museum of Antiquities in Leiden, Holland, a scribe named Ipuwer tells of a cataclysmic period that befell Egypt in the eighteenth or nineteenth century B.C.E. Ipuwer wrote about plague throughout the land, blood everywhere, fire, ruin, unbearable noise, and darkness, many of his comments hinting at distant volcanism and paralleling the later accounts in Exodus.

Most indicative of a volcanic eruption during the Exodus, however, is the biblical account of the flight itself: "And the Lord went before them by day in a pillar of a cloud, to lead them the way; and by night in a pillar of fire, to give them light" (Exodus 13:21). The pillars of fire and cloud probably refer to volcanic eruption clouds from somewhere in the eastern Mediterranean area, but the source is unknown. Only Thera is known to have had a major eruption in that region during the last two millennia, but it is unlikely that its fires could have been seen from Egypt.

The mythical island continent of Atlantis, the object of so much speculation over the centuries, most likely has its origins, too, in the Bronze Age eruption of Thera. The Athenian philosopher Plato, about 350 B.C.E., wrote two dialogues, *Timaeus* and *Critias*, in which he described the mythical island. Plato's account is the basis of all subsequent writing and speculation about Atlantis. He heard the story from Critias, who had heard it from his grandfather, whose father, named Dropides, had heard it from Solon (630?–560? B.C.E.), the great sage and lawgiver of Athens. In about 600 B.C.E. Solon had visited Saïs in Lower Egypt, where, according to Plato, Egyptian priests told him about a great island empire that had existed 9,000 years before and had vanished beneath the sea. In the Timaeus dialogue, Plato quotes Critias as relating the following, as told to Solon by one of the priests:

> there was an island situated in front of the straits which by you are called the Pillars of Heracles. The island . . . was the way to other islands, and from these you might pass to the whole of the opposite continent. . . . Now in this island of Atlantis there was a great and wonderful empire which had rule over the whole island and several others, and over parts of the continent. . . . But . . . there occurred violent earthquakes and floods, and in a single day and night of misfortune . . . the island of Atlantis . . . disappeared in the depths of the sea.[4]

The priest described a beautiful and fertile plain in the center of the island, and in the center of the plain was a hill on which lived a maiden named Clito. The sea god Poseidon fell in love with Clito and married her "and fortified the hill where she had her abode by a fence of alternate rings of sea and land, . . . one within another."[5] There were two springs on the island, one cold and one warm (most likely from volcanic activity), and there was an "abundance of food plants of all kinds." Poseidon and Clito had five pairs of twin sons and divided the island among them. The oldest, named Atlas, was made king over all the others, and the island was named Atlantis after him.

Atlantis was a place of peace and prosperity. There were magnificent temples, palaces, and fountains, and bridges over the circular canals, all made of stone. "The stone, black, white, and red, they quarried beneath the . . . central islet and outer and inner rings." Busy shipyards built great vessels called triremes, with three banks of oars, which were used in commerce with other lands. The Egyptian priest described that palace in the following passage, which could well be a description of the palace of Minos at Knossos: "This palace they originally built . . . in the dwelling place of the god . . . , and each monarch, as he inherited it in his turn, added beauties to its existing beauties . . . until they had made the residence a marvel for the size and splendor of its buildings."[6]

The time given by Plato for the demise of Atlantis, 9,000 years before Solon's visit to Saïs, or about 9600 B.C.E., must be wrong. No such advanced civilization could have existed on earth that long before the Bronze Age. Greek seismologist Angelos Galanopoulos has suggested that Plato's numbers are systematically off by a factor of ten.[7] Galanopoulos theorizes that Solon, instead of being *told* the story of Atlantis, read about it himself in ancient papyrus rolls made available to him in Saïs, and that he confused the written symbol for 100 with that for 1,000. Therefore Solon's 9,000 years should be 900, and thus the date should be 1400 B.C.E.—well within the Bronze Age. This date agrees quite well with the date suggested by archaeological studies, but it is two or three hundred years later than the date arrived at by scientific methods.

Plato's account implies that Atlantis was west of the Pillars of Heracles (or Hercules), which are traditionally assumed to border the Strait of Gibraltar. Thus Atlantis was long thought to have been in the Atlantic Ocean. The island was not named after the ocean, however—rather, both were named after Atlas—and literally hundreds of places around the world have been proposed over the years as the location of Atlantis.

In 1872 a French writer, Louis Figuier, suggested that Plato's Atlantis was the island of Santorini because parts of Santorini had obviously sunk into the sea. Indeed, on that archipelago today there are hot springs, as described by Plato, as well as black, red, and white volcanic rocks. And in 1885 a French archaeologist, Auguste Nicaise, proposed a correlation between the destruction of Atlantis and the Bronze Age eruption of Thera.

In 1909 the Irish scholar K. T. Frost, knowing about Evans's discoveries at Knossos and taking into account the extensive trade between Egypt and Minoan Crete, suggested that the Atlantis myth applied to Crete and the abrupt demise of Minoan civilization. To the Egyptians, Frost pointed out, the sudden cessation of economic contact with Crete would seem as if that island had disappeared into the sea. No doubt the Egyptians had heard about the destruction of Santorini from Minoan sailors, and they may have confused the two islands. In fact, the proximity of Crete and Thera, along with the rest of the Cyclades, makes that area the only location that provides the main elements of the Atlantis myth—an advanced civilization, an island empire, and a great natural catastrophe.

There can be little doubt that the story of Atlantis reflects Egyptian impressions of the highly evolved Minoan civilization that was based on Crete. And there can be little doubt that the Bronze Age eruption of Thera played a major role in the demise of that civilization, as reflected in ancient Egyptian texts. As we have seen, there seem to be ancestral memories of that eruption in several myths of ancient Greece and even, possibly, in certain events described in the Bible.

The Bronze Age weakening of Minoan culture enabled the Mycenaean Greeks to move into Crete, replacing the Minoan rulers at Knossos. Minoan émigrés from Crete infused their creativity and industry into the Mycenaean culture of Greece, thereby furthering Mycenaean expansion throughout the eastern Mediterranean area. And Minoan scribes at Knossos adapted their written language to accommodate the spoken language of their Greek conquerors, thereby developing the first, primitive form of written Greek.

Thus the Bronze Age eruption of Thera and the resulting Minoan diaspora appear to have been major factors in the rise of Mycenaean Greece, the cultural ancestor of classical Greece. The written language of the Mycenaeans evolved into the Greek script of the poets and philosophers of Athens during its Golden Age. And the Golden Age of Athens gave rise to the values and philosophies that underlie the culture we call Western civilization today.

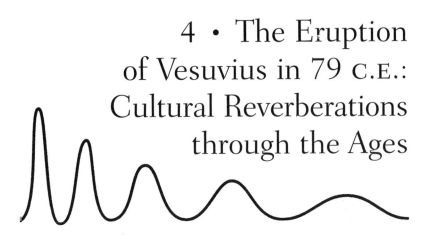

4 · The Eruption of Vesuvius in 79 C.E.: Cultural Reverberations through the Ages

No catastrophe has ever yielded so much pleasure to the rest of humanity as that which buried Pompeii and Herculaneum.

Johann Wolfgang von Goethe

MOUNT VESUVIUS, rising 1,279 meters above the Bay of Naples in southern Italy, is the only active volcanic mountain on the European mainland. Surrounded by cities and towns with an aggregate population of perhaps 3 million, it is both a magnificent landmark and an ever-present menace. When Vesuvius erupted in 79 C.E., it killed thousands of people, devastated the surrounding countryside, and destroyed at least eight towns, most notably Pompeii and Herculaneum—and it left a cultural and historical legacy that has resonated through Western civilization for almost 2,000 years.

Vesuvius today is the most widely known volcano on earth, and one of the most thoroughly studied. Its eruptions over the centuries not only have destroyed towns and killed a great many people but also have profoundly affected the economies of nearby areas for both better (rejuvenated soil and tourism) and worse (ruined farms and vineyards). In addition, the consequences of Vesuvius' activity are reflected in more than one branch of science. Indeed, the eruption of 79 C.E. can be said to have given birth to the science of volcanology and to have significantly advanced the science of archaeology. Moreover, Vesuvius has figured prominently in mythology, the arts, literature, even religion. It has become a part of the collective consciousness of the Western world.

The volcano, in solitary splendor, rises from the plain of Campania, a region of southern Italy between the Apennine mountains and the Tyrrhenian Sea. Blessed with a moderate climate and sunny skies, Campania is also a fertile land, its fertility enhanced by nutrients in the volcanic soil. In prehistoric times a cattle-herding people known as the Oscans settled in Campania and established villages. Greeks arrived by sea sometime around 800 B.C.E., attracted by the mild climate and verdant countryside. Gradually they assimilated the Oscans, and the culture of the area became predominantly Greek. Two hundred years later the Greeks lost political control to Samnite invaders from the interior, who in turn lost control to Rome in the Samnite Wars of the fourth century B.C.E.

The old Oscan villages became towns, and two of them grew into small cities—Pompeii and Herculaneum. Both were situated on the flanks of Vesuvius, Pompeii to the southeast and Herculaneum due west. In the first century B.C.E. Pompeii and Herculaneum took part in a regional rebellion. It was suppressed by the Roman dictator Sulla, who subsequently established a colony at Pompeii for veterans of his armies. Pompeii thrived thereafter as a Roman city, noted for wines, cabbages, and a fish sauce that was considered a great delicacy. Located near the seacoast on the north bank of the Sarno River, the city became an important center of trade with the interior. Its population grew to an estimated 20,000.

As Pompeii prospered, its wealthier citizens built luxurious houses, as did well-off Romans who desired a second home, or villa, in the salubrious climate of Campania. The villas— some in town, some in the countryside—were large, with open courtyards and many rooms. The interiors were beautifully decorated. Finely wrought statues of marble and bronze, most inspired by Greek mythology, were common in public rooms and courtyards, and floors in many houses were paved with intricate mosaic patterns. Walls were adorned with colorful frescoes, or they were richly painted with landscapes, fantastic architectural renderings, still lifes, contemporary scenes, and mythological images.

The Pompeiians had built a large, rectangular forum in their city, over 150 meters long and almost 40 meters wide, lined on both sides and at the south end with colonnades that formed covered walkways around the perimeter. On the west side of the forum was a basilica consisting of a long, colonnaded hall, or *nave* (from the Latin word for "ship"). Dating from about 120 B.C.E., the basilica of Pompeii is the earliest known example of a type of architecture that has been adapted over the centuries for a variety of public buildings, especially churches.

Pompeii also had two theaters, the largest of which could seat 5,000 people, and at least three public baths, all lavishly decorated. At the east end of the city was a huge, elliptical amphitheater where up to 20,000 spectators cheered as gladiators fought one another, and wild animals, to the death. That amphitheater, which dates from 70 B.C.E., antedates the Colosseum in Rome by as much as a century and a half and is among the oldest elliptical structures known.

The historical importance of Pompeii, however, lies not so much in its public buildings as in the many well-preserved buildings used by people in their day-to-day lives. The remains of more-impressive temples and forums can be found elsewhere in the Mediterranean region, but there are no better examples of the homes and shops of a city's ordinary citizens.

Herculaneum, with a population estimated at 4,000 to 5,000, was considerably smaller than Pompeii. Located on the coast about 15 kilometers northwest of Pompeii, it was both a resort where, as in Pompeii, wealthy Romans built villas, and a center of commercial fishing and possibly shipbuilding. Like Pompeii, it boasted a forum, a theater, and public baths. Herculaneum's baths were, in fact, considerably more sumptuous than Pompeii's. The public buildings and private houses in

Herculaneum, like those in Pompeii, were profusely decorated with works of art. Paintings and frescoes that have been found in both cities, as well as numerous graffiti, especially in Pompeii, reveal much about the society, politics, dress, vocations, and culture of the people who inhabited that part of Italy in Roman times.

———

Mount Vesuvius developed inside the caldera of an older volcano, Monte Somma. Only the northern flank of Monte Somma remains, forming a wall-like ridge around the north side of Vesuvius (Figure 4-1). In the last four millennia, the Somma-Vesuvius complex has been the source of four major eruptions (1550 B.C.E. and 79, 472, and 1631 C.E.) that had a regional impact. In the intervening periods as many as fifty more-localized events occurred.

Vesuvius is located in the Romana volcanic belt, which extends some 450 kilometers from Mount Amiata, about 130 kilometers northwest of Rome, to Mount Vulture, about 110 kilometers east of Naples (Figure 4-2). The Romana belt is composed of two parallel volcanic arcs. The older, eastern arc is probably extinct. The western arc, which includes Vesuvius, is only a few hundred thousand years old.

Both arcs are convex toward the southwest. Ordinarily the convexity of a volcanic arc implies that the tectonic plate on the convex side of the arc is sliding, or being subducted, beneath the plate on which the arc developed. Thus it would seem that the Tyrrhenian platelet, west of Italy, is being subducted beneath the Italian peninsula to the east. The focuses of earthquakes in that region, however, become deeper toward the west, indicating that a zone of tectonic activity dips westward, not eastward. Therefore instead of the Tyrrhenian platelet being subducted toward the east, it may be that the Apulian platelet, east of Italy, is slipping westward beneath both Italy and the Tyrrhenian Sea.

Or perhaps it would be more accurate to say that the Tyrrhenian platelet is being thrust over the Apulian platelet. If

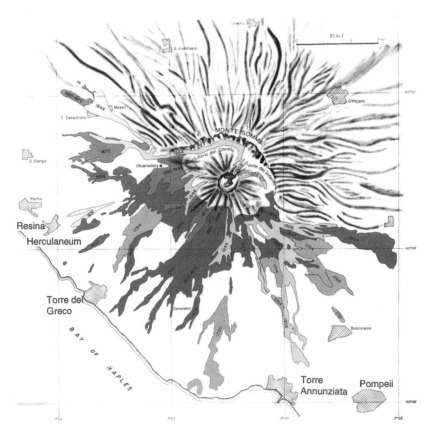

FIGURE 4-1. Bird's-eye view of Vesuvius and Monte Somma showing various lava flows, most emplaced during the past three centuries. From Bullard, *Volcanoes of the Earth*.

indeed that is the case, present-day volcanism in the Romana belt cannot be attributed to the vertical rising of molten rock (magma) from a wedge of mantle material above the subduction zone, the usual process. Instead, the mantle wedge itself must be working its way upward along the westward-dipping Apulian platelet, moving toward the zone that separates that platelet from the Tyrrhenian platelet. It is from that highly fractured region that molten rock presumably rises into magma chambers beneath the volcanic complexes of western Italy.

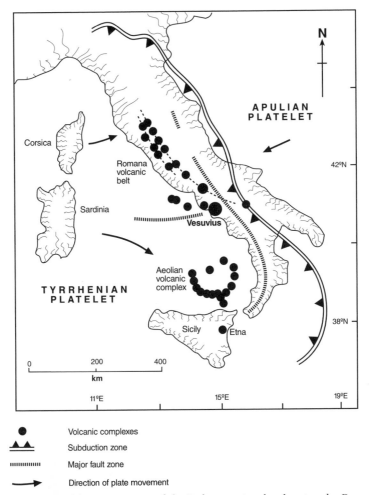

FIGURE 4-2. Tectonic setting of the Italian peninsula, showing the Romana volcanic belt, which originated above the collision zone between the Apulian and Tyrrhenian platelets. The Aeolian volcanic complex, to the south, is believed to have originated above the collision zone between the African plate and the Tyrrhenian platelet.

Geophysical studies show that the magma chamber beneath Vesuvius is some 5 kilometers below the volcano, almost 2 kilometers high, and about a kilometer in diameter. Those studies also indicate that Vesuvius formed at the intersection of two major fractures in the earth's crust—a northwest-trending fault similar to many in the Apennines and a northeast-trending fault that extends beneath the Bay of Naples.

It was long assumed that Vesuvius had not been active in historical times prior to the eruption of 79 C.E. But in 217 B.C.E. there were violent earthquakes in Italy, and there were reports of a haze, or dry fog, that dimmed the sun. The Greek biographer Plutarch reported "a sky on fire" near Naples, and the Roman poet Silius Italicus, in an account of "prodigies" during the year 217, wrote, "Vesuvius also thundered, hurling flames worthy of Etna from her cliffs; and the fiery crest, throwing rocks up to the clouds, reached to the trembling stars."[1] Those accounts are supported by high levels of acidity found in cores of Greenland ice dating back to that time. The acidity—which is assumed to have come from atmospheric hydrogen sulfide emitted by an erupting volcano—provides strong evidence that Vesuvius was indeed active some 300 years before 79 C.E.

If there *was* an eruption in 217 B.C.E., the people of Campania apparently had forgotten about it by the year 79. They thought of Vesuvius as a benign mountain, one that picturesquely dominated the beautiful Bay of Naples (then called Neapolis) and whose slopes were planted in lush vineyards. With few exceptions, no one considered the mountain volcanic, like the oft-erupting Etna in Sicily or the fiery Stromboli in the Tyrrhenian Sea. The shallow depression at the summit of Vesuvius was not recognized as a crater. Indeed, in 72 B.C.E. the gladiator Spartacus and his rebellious followers sought refuge there from their Roman pursuers.

Several years earlier, however, the Greek geographer Strabo (63 B.C.E.?–24 C.E.?) had written of Vesuvius, describing its summit in these words: "all of it is unfruitful, and looks ash-coloured, and it shows pore-like cavities in masses of rock . . . looking as though they had been eaten out by fire; and hence one might infer that in earlier times this district . . . had craters of fire."[2] Strabo's observations were largely ignored, but in 79 C.E., half a century after his death, his inference of volcanism was tragically confirmed.

Despite their ignorance of the nature of Vesuvius, the people of Campania were not unacquainted with volcanic phenomena. A short distance west of Naples, an area called the Phlegraean Fields (see Figure 4-3) was—and still is—pockmarked by craters, some now containing lakes and others containing boiling mudpots and vents called fumaroles, which emit hot gases. One of the craters, La Solfatara, named by the ancients for its volcanic sulfur deposits and sulfurous emissions, was thought to be the forge of Vulcan, the Roman god of fire and metalworking. Geologists now use the term *solfatara* for all sulfurous fumaroles. Another crater, which today

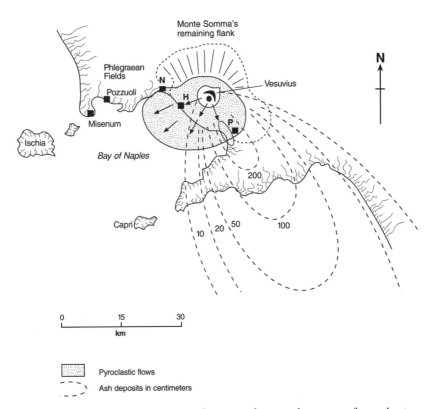

FIGURE 4-3. Mount Vesuvius and vicinity, showing the extent of pyroclastic flows and ash deposits from the eruption of 79 C.E. H: Herculaneum, P: Pompeii, N: Naples. Adapted from Sigurdsson et al., "Eruption of Vesuvius."

contains Lake Avernus, figured in the *Aeneid* of the Roman poet Virgil as the entrance to the underworld. Covering some 65 square kilometers, the Phlegraean Fields are located in a caldera that formed about 12,000 years ago. That caldera, in turn, lies within a larger caldera, some 11 kilometers across, that formed about 37,000 years ago.

The Italian poet Dante Alighieri (1265–1321), though not known to have visited the Phlegraean Fields or Vesuvius, was quite likely inspired by reports of that volcanic area. His *Inferno,* a fantastic account of a visit, with Virgil, to the nine levels of hell, incorporates many features that resemble the Phlegraean fumaroles and boiling mudpots—the burning tombs of heretics, for example, and the burning holes into which simonists (dealers in ecclesiastical favors) are plunged headfirst. And the stench of Dante's hell evokes the sulfurous reek of La Solfatara. Moreover, Dante has blasphemers, sodomites, and usurers doomed to walk forever on burning sands in a continuous rain of fire—a passable description of the fiery ash that spewed from Vesuvius in Dante's time, as it had often done before and has done many times since.

Near the center of the Phlegraean Fields caldera is the fishing village of Pozzuoli (see Figure 4-3). In Roman times the village was called Puteoli after the putrid effluvia of the nearby fumaroles. At the bottom of the sea a short distance up the coast, and visible today from the air, lie the ruins of an even older town known as Port Julius, its history lost in the mists of time. On the shore at Pozzuoli are the ruins of a first-century market. The ruins include standing marble pillars long thought to be the remains of a temple honoring the Roman god Serapis. The ruins have been alternately submerged and exposed over the centuries because the ground in and around Pozzuoli has been subjected to uplift and subsidence as magma intrudes beneath the caldera and then moves elsewhere. Similar subsidence in ancient times was responsible for the submergence of Port Julius.

In 1830 the British geologist Charles Lyell used a lithograph titled the "Temple of Serapis" as the frontispiece in his seminal book *Principles of Geology*. That pioneering work widely publicized the theory of uniformitarianism, the concept that the present is the key to the past—that is, that processes at work on earth today, acting continuously over millions of years, can explain past geological phenomena. The columns in the ruins at Pozzuoli, when uplifted and exposed, reveal bands where marine organisms have bored into the marble. The bands demonstrate that those parts of the columns have been beneath the sea in the past—proof that land can subside and be uplifted, just as geological evidence shows that land masses were subjected to repeated subsidence and uplift long ago.

The restless magma beneath the Phlegraean Fields has made the entire area earthquake prone. Over the years quakes have caused considerable damage in Pozzuoli. Many of the town's residents fled when earth movements intensified in recent years, but most have returned, tempting fate.

In 79 C.E., of course, nobody knew about plate tectonics and the fact that earthquakes can be harbingers of volcanic activity. Earthquakes were common throughout Campania, but they were explained by myths that recounted prodigious battles between the gods and a race of giants. The gods eventually prevailed and imprisoned the giants in the underworld, where their struggles to free themselves shook the earth.

One writer, the Roman historian Dio Cassius, did link the two phenomena, in a way. Discussing the eruption of Vesuvius more than a century after 79 C.E., he wrote,

> Numbers of huge men quite surpassing any human stature—such creatures, in fact, as the Giants are pictured to have been—appeared, now on the mountain, now in the surrounding country. . . . Then suddenly a portentous crash was heard, as if the mountain were tumbling in ruins; and first huge stones were hurled aloft, . . . then came a great quantity of fire and endless smoke, so that . . . day was turned into night. . . . Some thought that the Giants were rising again in revolt (for at this time also many of their forms could be discerned in the smoke . . .).[3]

And Virgil, writing long before the eruption, speculated that one of the defeated giants was buried beneath Vesuvius. Nobody, however, extrapolated from mythology to a causal connection between volcanoes and earthquakes.

Thus when a powerful earthquake shook Campania in February of 62 C.E., nobody thought it might herald a volcanic eruption. The quake caused great damage, especially in Pompeii and Herculaneum. Pavement cracked, walls crumbled, roofs collapsed, and columns toppled, and in Pompeii, streets were flooded when the city's water reservoir gave way. There were many injuries and deaths. The Roman philosopher Seneca wrote that a flock of 600 sheep died near Pompeii. He attributed their deaths to a pestilence caused, he believed, by poisons from within the earth. Seneca's description suggests that the earthquake opened fissures through which volcanic gases reached the surface.

Another earthquake the following year damaged Neapolis (now Naples) while the emperor Nero, who flattered himself that he was a gifted singer, was giving a concert. It is said that he sang on, even though the theater was shaking. In Rome, the Senate authorized funds for reconstruction in the region. The citizens of Pompeii and Herculaneum were still rebuilding when the knock-out blow came from Vesuvius in the year 79.

On August 24, 79 C.E., the people of Pompeii and Herculaneum were going about their everyday activities—taking care of domestic matters at home, conducting business in the forum, socializing in the baths, quaffing wine in taverns, or perhaps paying homage in temples dedicated to their favorite gods—when Vesuvius exploded with a deafening roar (Figure 4-4) and turned their benign, well-ordered world into a hellish nightmare. The earth shook, and ash, frothy bits of pumice, and sand-size particles called *lapilli* shot high into the atmosphere in a dense column. The top of the column reached the stratosphere and spread out to form what in today's atomic age would be called a mushroom cloud. That initial phase of the eruption continued for almost twelve hours. It had an

FIGURE 4-4. Eruption of Vesuvius in 1779, as seen from Naples. The painting shows what the initial eruption of 79 C.E. may have looked like. From
Alfano and Friedlaender, *Die Geschichte des Vesuv,* plate 24.

estimated VEI (volcanic explosivity index) of 6, categorized as "huge" by volcanologists.

Pompeii was downwind from Vesuvius, about 10 kilometers from its crater. About half an hour after the start of the eruption, ash and pumice began falling on the city at a rate of 15 to 20 centimeters an hour, like a diabolical snowstorm, and blotted out the light of the sun. People took shelter in buildings or else fled in terror through darkened streets. Many were killed by asphyxiation. By late afternoon, roofs were caving in from the weight of accumulated debris, killing more. By midnight the fallout was two and a half meters deep in places. The streets of Pompeii were buried, and the dead were entombed where they fell. The finer particles of ash and dust at the top of the eruption column drifted southward with the prevailing winds, eventually reaching northern Africa.

At about midnight, decreasing pressure in the magma chamber beneath the volcano caused the eruption column to collapse, and a second phase began. Starting shortly after midnight on the twenty-fifth, the volcano produced fiery avalanches of hot gases, ash, pumice, and lapilli—pyroclastic flows—that roared down the western slopes of Vesuvius (Figure 4-5), igniting vegetation and habitations in its path and smashing into Herculaneum (see Figure 4-3). A pyroclastic flow of the type produced by Vesuvius typically divides into two parts: a turbulent cloud of fine debris and fiery volcanic gases, called a surge, which moves very rapidly, and a slower-moving ground-hugging flow of denser material, including sizable fragments of rock.

The surge, probably with a temperature in the range of 100 degrees Celsius, reached Herculaneum first. Traveling at speeds of up to 300 kilometers an hour, it ripped roof tiles from buildings, toppled columns and statues, covered the streets, and killed virtually everyone who had not fled when the eruption began. The heavier flow moved more slowly,

FIGURE 4-5. Eruption of Vesuvius in 1810, as seen from Naples. The painting may depict a pyroclastic flow like those produced by Vesuvius in 79 C.E. during the later phases of the eruption. From Alfano and Friedlaender, *Die Geschichte des Vesuv,* plate 35.

but its temperature was as high as 400 degrees. It burned or charred all flammable objects—wooden timbers, parts of human bodies—that had not been covered by the surge deposits, and it sealed the ruins.

Scholars long assumed that Herculaneum had been buried by volcanic mudflows rather than by the kind of loose ash and pumice that buried Pompeii, for at Herculaneum the deposits had become almost rock-hard, as mud would be expected to do over time. But the consensus today is that the material that inundated Herculaneum came from a series of pyroclastic flows, which also produce deposits that are more compact and consolidated than ash and pumice.

The first pyroclastic flow did not reach Pompeii, nor did a second. Later that morning the fiery surge from a third flow reached the northern walls of Pompeii but apparently did no great damage. The following flow of heavier debris, however, buried what remained of the city.

People in Pompeii who had stayed indoors during the first phase of the eruption or had managed to get under cover survived longer than those caught in the open, unless they were killed by collapsing roofs. But the rapidly accumulating volcanic debris trapped many in their homes, where they died from asphyxiation when a fourth pyroclastic flow from Vesuvius, the morning of August 25, engulfed the city. It knocked down any walls that protruded above the city's volcanic blanket.

Vesuvius produced two more pyroclastic flows on August 25, each larger than those preceding. The final blast was accompanied by strong earthquakes and a gigantic black cloud, which darkened the Bay of Naples and the countryside for miles around. But by that time both Pompeii and Herculaneum were dead and buried. They would not be seen again for almost 2,000 years. And all the countryside on the western and southern flanks of Vesuvius was devastated. Farms, vineyards, villas—all were destroyed.

Ash and pumice continued falling during the intervals between pyroclastic surges. Figure 4-6 indicates the various surges that struck Pompeii and the thicknesses of the intervening ash and pumice deposits at one excavated site within the city. Much of Pompeii ultimately was covered by almost 5 meters of volcanic debris. Herculaneum, closer to the volcano, was buried as much as 20 meters deep. The cumulative volume of volcanic material has been estimated at about 8 cubic kilometers. Ash covered some 500 square kilometers of land.

There are indications that a few survivors returned to Pompeii and tunneled into the ruins in an effort to retrieve valuables, and no doubt thieves sought plunder. In the loose ash, however, the tunnels collapsed and were soon abandoned. In

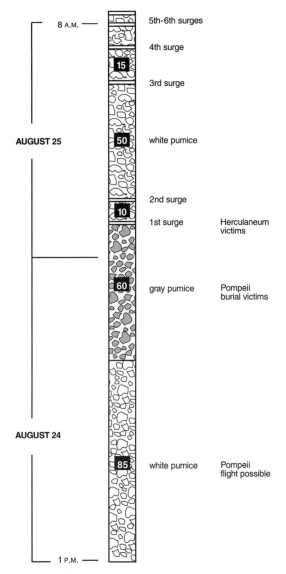

FIGURE 4-6. Chronology and thicknesses of volcanic products that fell on Pompeii during the intervals between pyroclastic surges during the eruption of 79 C.E. The numbers in black squares indicate thicknesses in centimeters. Adapted from Sigurdsson et al., "Eruption of Vesuvius," 346.

any event, most of the former residents were dead or had resettled elsewhere. As the years went by, both Pompeii and Herculaneum were all but forgotten.

Indeed, only one contemporary account of the disaster exists. A Roman naval officer named Gaius Plinius Secundus, known to history as Pliny the Elder, author of a multivolume treatise on natural history, was in charge of a Roman naval squadron based at Misenum, on the peninsula that forms the northern side of the Bay of Naples. When the eruption started, Pliny commandeered one of his galleys and started off in the direction of Vesuvius, hoping to get a close-up view of the eruption and also rescue a friend who lived on the coast near the mountain. He was driven away, however, by hot ash and lapilli, and he changed course for Stabiae, on the south side of the bay, where another friend had a villa. It was there, while awaiting favorable conditions for the return to Misenum, that Pliny died, probably of a heart attack, while struggling through a heavy ashfall that buried the town.* Some years later Pliny's nephew, known today as Pliny the Younger, was asked by the Roman historian Tacitus for an account of his uncle's death.

The younger Pliny, who was but seventeen or eighteen years old when the eruption occurred, had been living with his uncle at Misenum. He wrote two letters to Tacitus describing not only how his uncle had died, but, in some detail, the eruption itself and an accompanying earthquake. Those letters contain the earliest known eyewitness account of a volcanic event. In the first letter, Pliny described the initial eruption column, with its mushroom cloud at the top, as "being like an umbrella pine, for it rose to a great height on a sort of trunk and then split off into branches."[4] Volcanologists now use the Italian word *pino* to describe such eruption columns, and highly explosive eruptions like that of Vesuvius on August 24 are called *Plinian*.

---

*Like Pompeii and Herculaneum, Stabiae remained buried for centuries. Today the site is occupied by the city of Castellammare di Stabia.

In his second letter to Tacitus, Pliny described a startling phenomenon in the Bay of Naples. "We . . . saw the sea sucked away," he wrote, "and apparently forced back by the earthquake: at any rate it receded from the shore so that quantities of sea creatures were left stranded on dry sand."[5] That change in sea level most likely was caused not by an earthquake but by a tsunami—a great sea wave—created by a pyroclastic flow when it slammed into the Bay of Naples (see Figure 4-3). The sea commonly withdraws from the shore just before a tsunami strikes.

———————

Paradoxically, the cataclysm that destroyed Pompeii and Herculaneum saved those cities from oblivion. Though in ruins, they were preserved, like insects in amber, for discovery centuries later. Most cities of antiquity have been irretrievably lost—obliterated in warfare, buried beneath subsequent construction, or simply crumbled to dust. The material ejected from Vesuvius did not obliterate Pompeii and Herculaneum; it entombed and protected what was left of them. Streets, walls of houses and public buildings, magnificent works of art, and human remains were preserved relatively intact—like a snapshot of Roman life in 79 C.E. Pompeii today is among the largest such sites known from the ancient world. No museum, however vast, could recreate such glorious windows into an ancient civilization. Pompeii and Herculaneum are places where the past is revealed almost in living immediacy.

The discovery of the two cities was a fitful process. The first hint came in the late 1500s, when a canal was being dug to bring water from the Sarno River to a village a short distance to the north. Workmen uncovered some fragments of marble and a few coins dating from the reign of Nero, but nobody connected the finds to ancient Pompeii. Then in 1689 the digging of a well in the area uncovered a set of iron keys and some stones bearing Latin inscriptions, one of which spelled out "Pompeii"—but it was assumed that the items had come from a country villa.

Curiously, an expanse of slightly elevated, open land just north of the Sarno was known to local peasants as la Città, "the city." An ancestral memory, perhaps? In any event, an investigator named Giuseppe Macrini tunneled into la Città and, in a book published in 1699, claimed to have seen evidence of houses and city walls. Macrini had found Pompeii, but, incredibly, nobody seems to have paid any attention.

The scene shifts now to Herculaneum. In 1709 Prince d'Elboeuf, an officer in the Austrian army, which at that time occupied the Kingdom of Naples, was building a villa at nearby Portici. He heard that some pieces of carved marble had been found in a well or spring in the area. D'Elboeuf hired workmen to sink a shaft, hoping to find the source of the marble fragments. What he found were the remains of Herculaneum's theater, which, through tunnels, he plundered for bronze and marble statues to grace his villa.

D'Elboeuf neither knew nor cared that his treasure trove was part of ancient Herculaneum. A few years later he sold his villa, which eventually was acquired by the Bourbon King Charles of Naples and Sicily. Tunneling in the area continued, and priceless statuary and slabs of marble were carted away to adorn the villas of the nobility. Then in 1738 an inscription was found that included the name of the city. But the interest was in treasure hunting, not history. Tunneling and excavation continued randomly, the digging sites changing whenever salvageable treasures were not found immediately.

Interest in la Città revived after the discovery of Herculaneum, and excavation was renewed there. In 1763 workmen found part of a statue bearing the name Pompeii, and finally it was realized that la Città was indeed the site of the lost city. Digging continued apace, but still chaotically. Pompeii and Herculaneum were regarded merely as quarries—sources of artworks for the wealthy—not as sites to be studied for their historic value.

Although it has been said that the science of archaeology began with the discoveries of Herculaneum and Pompeii, its systematic application had to wait another hundred years. After the unification of Italy in 1860, an archaeologist named Giuseppe Fiorelli was named superintendent of all excavations in southern Italy. Under Fiorelli, the digging, for the first time, was carried out more systematically. In Pompeii, for example, he divided the city into areas of reference, logged each find in a journal, and even founded a school of archaeology.

Excavation at Pompeii has proceeded much more rapidly than at Herculaneum. Pompeii was buried beneath dry, unconsolidated ash, pumice, and lapilli—materials that have been relatively easy to remove. At Herculaneum the digging has been more difficult because the overlying deposits, having originated in pyroclastic flows, are much more consolidated. Moreover, Herculaneum was buried more deeply than Pompeii. The modern town of Ercolano lies almost 20 meters above a large part of the buried city, so much of Herculaneum is unreachable except through tunnels. Pompeii is mostly covered by farmland, however, and today less than a quarter of the city remains to be uncovered.

No doubt a number of people escaped from both Pompeii and Herculaneum during the first hours of the eruption, but probably few survived the subsequent pyroclastic flows. It was long thought, judging by the number of human remains found, that only about 2,000 people died in Pompeii. Recent estimates, however, put the toll much higher, possibly as high as 16,000—especially when probable deaths in the surrounding countryside are figured in.

The excavations in Pompeii have revealed human remains in two quite different forms. On the one hand, victims who were trapped within their houses—either because they thought they could wait out the initial ashfall or possibly because they were reluctant to leave their homes and possessions—were killed by collapsing roofs or asphyxiated by pyroclastic flows and subsequently covered with ash and lapilli. Water and air eventually penetrated those loosely packed materials, and soft human tissue, as well as clothing, wood, and other perishables, decayed over time. Their skeletons have been found where they died, often crouched in corners, vainly seeking shelter from the terrible fate that engulfed them. Perfectly preserved jewelry adorned some of the skeletons when they were found, and near some of the bones lay coins spilled from bags that had long since rotted away.

On the other hand, those Pompeiians who perished in the streets, suffocated by falling ash, left not skeletons but molds of their bodies. The ash became packed around them as they fell. In time it became damp from groundwater seepage and, like clay, held its shape as the bodies decayed. In the 1860s Giuseppe Fiorelli discovered that by pouring liquid plaster into the molds thus formed, he could produce casts of the victims as they had looked at the moment of death—the attitudes of their bodies, imprints of their clothing, and, in some cases, the expressions on their faces. The results are awesome, even eerie, and profoundly moving.

In Herculaneum there are no body molds like those found in Pompeii because the initial ashfall did not reach that city. And only about a dozen human skeletons had been found in the excavated ruins until 1982, when workers digging a drainage ditch in the area that had been the city's waterfront unearthed a series of arches that supported buildings above them. The chambers formed by the arches were used by fishermen to store boats and other equipment. Found within the chambers were many skeletons of people who no doubt had fled there when the eruption began, in hopes of getting away by boat. But perhaps there were not enough boats, or perhaps the sea was too rough. In any case, the refugees died when the first pyroclastic flow struck the city.

The skeletons found in Herculaneum are in a much better state of preservation than those found in Pompeii. Groundwater percolating through the volcanic debris covering the bones sealed them against oxidation and protected them from changes in temperature and humidity. Analyses of the bones found under the arches have enabled anthropologists to determine the age, state of health, and kind of work done by each person. Most were healthy and well fed. Examination of points where muscles were attached to bone has shown that some, no doubt slaves, had done heavy work. Others, presumably wealthy members of the aristocracy, had done little physical labor. Perfectly preserved jewelry, some quite magnificent, found in association with some of the skeletons also testifies to their social status in life. Thus men, women, children, aristocrats, slaves—all perished together, in a common, democratic grave.

For the science of physical anthropology, the skeletons of Herculaneum were a rare find indeed. The Romans usually cremated their dead, so anthropologists have had little direct information about the people themselves. The same conditions that preserved the skeletons also preserved artifacts that, in Pompeii, would have long since decayed. Utensils, wooden furniture, fishnets, even foodstuffs, though charred, have been found much as they were on that fateful August day in 79 C.E. Thus the ruins of Herculaneum have yielded a wealth of new information about ancient Rome.

———

The science of volcanology can be said to have its roots in Mount Vesuvius. Figure 4-7, from the *Encyclopédie* of the French philosopher Denis Diderot (1713–1784), shows people fascinated by a close-up view of an eruption of Vesuvius in 1754. This early evidence of intellectual curiosity may well have heralded later scientific interest in volcanism. Sir William Hamilton, the British envoy in Naples from 1764 until 1800, witnessed several eruptions of Vesuvius and became fascinated by the mountain. He was an excellent observer and an avid student of the volcano, making many trips to the summit, even during eruptions. His observations, though only semiscientific by today's standards, were set forth in numerous letters that were published in the *Philosophical Transactions* of the Royal

FIGURE **4-7**. A figure from the late-eighteenth-century *Encyclopédie* compiled by the French philosopher Denis Diderot, showing an eruption of Vesuvius in 1754 and early public interest in viewing such phenomena close at hand. This emerging curiosity may well have heralded later scientific interest in volcanism. Private collection.

Society of London. In addition, Hamilton wrote two books on volcanic phenomena. His *Observations on Mount Vesuvius, Mount Etna, and Other Volcanoes* was published in 1774, and two years later he published *Campi Phlegraei* (Phlegraean Fields). Hamilton's work constituted the first modern writings on volcanology, and, though he was a diplomat not a scientist, he rightly has been called "the father of volcanology."

A singular advance in the science was the construction of an observatory, completed in 1845, on a ridge high on the northwest flank of Vesuvius. Built solidly of stone, it was equipped with seismographs and other instruments, and it contained a library and living quarters for a director. In 1856 Luigi Palmieri, a physicist from Naples University, was appointed director of the observatory, and in 1872 he courageously stayed there through a major eruption. Palmieri determined, over the years, that the eruptions of Vesuvius seemed to follow cyclical patterns; his was the first effort to predict the behavior of a volcano.

Vesuvius has erupted more than fifty times since 79 C.E. There were especially destructive eruptions in 472, 512, and 1139, and a disastrous eruption, with pyroclastic flows, occurred in 1631. Watching from a nearby monastery, one Fra Angelo reported seeing seven streams of lava flowing toward the Bay of Naples from fissures in the southwestern flank of the mountain. Fra Angelo's "lava flows" were actually pyroclastic flows. Three towns—Granasello, Resina, and Torre del Greco—were severely damaged. As many as 18,000 people are believed to have been killed, most of them in Torre del Greco because local officials hesitated to order the city evacuated. When the evacuation order finally came, it was too late. Before the order could be carried out, pyroclastic flows traveling at express-train speed burst through and over the town walls and into narrow streets choked with panicked citizens.

In the village of Portici, after the eruption of 1631, a marble plaque was erected with a caveat warning the citizens that Vesuvius was a great danger to them. The inscription reads as follows:

> Posterity, posterity, this is your concern. . . .
> Be attentive.
> Twenty times, since the creation of the sun
> has Vesuvius blazed, never without a horrid
> destruction of those that hesitated to fly.
> This is a warning, that it may never
> seize you unapprized.
> The womb of this mountain is pregnant with
> bitumen, alum, iron, gold, silver, nitre,

and fountains of water.
Sooner or later it kindles. . . .
If you are wise, hear this speaking stone.
Neglect your domestic concerns, neglect your
goods, and chattels, there is no delaying.
Fly.[6]

There is a vivid eyewitness account of another major eruption, in October 1767, which a Catholic priest, Padre Torre, described this way:

> on a sudden, about noon, I heard a violent noise within the mountain, and at a spot about a quarter of a mile off the place where I stood the mountain split; and with much noise . . . a fountain of liquid fire shot up many feet high, and then like a torrent rolled on directly towards us. . . . in an instant clouds of black smoke and ashes caused almost a total darkness; the explosions . . . were much louder than any thunder I ever heard, and the smell of the sulphur was very offensive.[7]

One reason why the early eruptions of Vesuvius are so well documented is that Roman Catholic priests in the Cathedral of Naples have invoked protection from their patron, Saint Januarius, or San Gennaro, whenever the volcano has threatened the city (see Figure 4-8). Januarius, the bishop of Beneventum, was martyred at Puteoli (now Pozzuoli) around 300 C.E. during the persecution of Christians by the emperor Diocletian. Holy relics—the saint's skull and vials allegedly containing samples of his blood—have been ceremoniously paraded through the streets whenever Vesuvius has erupted, in hopes that San Gennaro will intercede to protect the city. Each of those occasions has been duly entered in diocesan records over the centuries.

When Padre Torre reached Naples after the 1767 eruption he found that city in chaos. He wrote,

> the churches were filled; the streets were thronged with processions of saints, and various ceremonies were performed to quell the fury of the mountain.

FIGURE 4-8. An old lithograph depicting the 1631 eruption of Vesuvius, showing Saint Januarius (San Gennaro) imploring the volcano to cease its destructive activity. Private collection.

In the night of the 20th, . . . the mob set fire to the gates of the Cardinal Archbishop because he refused to bring out the relics of St. Januarius. The 21st was a quieter day, but the whole violence of the eruption returned on the 22nd. . . . Ashes fell in abundance in the streets of Naples. . . .

In the midst of these horrors, the mob, growing tumultuous and impatient, obliged the Cardinal to bring out the head of Saint Januarius . . . ; and it is well attested here that the eruption ceased the moment the saint came in sight of the mountain.[8]

A heroic statue of San Gennaro, with his arms upraised, can be found today on the bridge between Naples and San Giorgio. Many Neapolitans believe the statue originally stood with its arms at its sides and that San Gennaro raised the arms toward the volcano during the night of April 30, 1832, when Vesuvius erupted and lava flowed. Soon the lava flows stopped, and Naples was saved. Belief in the saint's powers remains strong in Campania, and still today his relics are ceremoniously paraded through the streets of Naples whenever Vesuvius shows signs of activity.

The volcano was last active in March 1944, during World War II, when a powerful eruption devastated the towns of San Sebastiano and Massa and damaged an American air base. The Italian writer Curzio Malaparte wrote about the event in his book *The Skin:*

> The sky . . . was scarred by a huge, crimson gash, which tinged the sea blood-red. The horizon was crumbling away, plunged headlong into an abyss of fire. . . . the earth trembled, the houses rocked on their foundations. . . . A dreadful grinding noise filled the air. . . . And above . . . the wails and terrified shrieks of the people, who were running hither and thither, . . . there arose a terrible cry, which rent the heavens.
>
> Vesuvius was screaming in the night, spitting blood and fire. . . . A gigantic pillar of fire rose sky-high. . . . Down the slopes of Vesuvius flowed rivers of lava, sweeping toward the villages which lay scattered amid the green of the vineyards.[9]

Since 1944, Mount Vesuvius, ominously, has been in its longest period of repose in recent history.

———

Ever since Pompeii and Herculaneum were unearthed, those ancient cities, and Vesuvius itself, have had far-reaching significance in Western culture. Vesuvius is the single most depicted volcano in Western art. In eruption and in repose, it has been a favorite subject of painters, including Edgar Degas and Pierre-Auguste Renoir of France, Pieter Bruegel of the Netherlands, England's William Turner, and the Americans Albert Bierstadt and Thomas Cole. Even "pop" artist Andy Warhol painted the mountain. Many artists, notably Cole, the

Italians Giambattista Piranesi and Giovanni Cipriani, and Angelica Kauffmann, born in Switzerland, have depicted scenes from Pompeii and Herculaneum. Kauffmann's *The Younger Pliny and His Mother at Misenum* depicts the eruption of Vesuvius as Pliny must have seen it.

So, too, the works of art found in Pompeii and Herculaneum have fascinated connoisseurs and historians since those Roman cities were uncovered. Very little artwork has survived from ancient Rome itself, that imperial city having been sacked and plundered by the Vandals, the Ostrogoths, and others over the centuries. Thus the artists who decorated houses and public buildings in the buried cities have given us the most important evidence we have about artistic techniques in Roman times. Those artists covered interior walls with landscapes— indeed, they can be said to have invented landscape painting— which realistically imitated nature and created an illusion of spaciousness. Paintings, frescoes, and mosaics depicted scenes from Roman and Greek mythology and also scenes of people going about everyday tasks—scenes that are invaluable for understanding how ordinary people looked and lived during the first century C.E. Many sculptures were copies of now-lost Greek and Roman originals. There were exquisitely worked pieces of jewelry, cameos, drinking cups, lamps, candelabra, and pieces of household furniture.

The works of art uncovered in Pompeii and Herculaneum, and the building styles preserved in those cities, set the tone for the eighteenth-century rise of neoclassicism in the decorative arts. They emphasized order and simplicity, as contrasted with the elaborate ornamentation of the earlier, rococo style. Thus eighteenth- and nineteenth-century architecture, household furniture, ceramics, jewelry, textiles, and even wallpaper reflected Pompeiian motifs.

The neoclassical work of Scottish architect Robert Adam, especially, was inspired by the Pompeiian style. "Pompeiian rooms," with ceiling decorations, wall panels, medallions, and plasterwork incorporating Pompeiian motifs, were constructed in the homes of the wealthy. One such room designed by Adam, the drawing room of Lansdowne House in England, was dismantled, brought to the United States, and is now a permanent exhibit at the Philadelphia Museum of Art. And objects excavated from Pompeii and Herculaneum have provided models for a wide variety of decorative pieces, ranging from candelabra to plant stands. The decorative arts continued to be influenced by Pompeii well into the nineteenth century and, to a somewhat lesser extent, into the twentieth.

Among the Pompeii-inspired ornaments in Lansdowne House were plaques bearing bas-reliefs of dancing figures. The first marquis of Lansdowne, while touring Italy, bought a number of plaster bas-reliefs from street vendors. Upon returning to England he showed them to his friend Josiah Wedgwood, the English potter, who had molds made from them. Wedgwood adapted the dancing figures for use in his earthenware and stoneware pottery, as well as in wall plaques and sconces and even in small medallions used in ornamenting furniture. Those designs remain popular to this day.

Sir William Hamilton played a key role in introducing the public to Pompeiian art, especially in Great Britain. A notable collector of antiquities, Hamilton acquired (not always legally) many works of art, especially vases and statuary, from the diggings at Pompeii. He then sold them to wealthy collectors or to the British Museum in London, where they generated great interest and did much to sustain the popular enthusiasm for Pompeiian, hence Roman, style, not only in England but also in the United States and elsewhere.

At least two notable buildings in the United States are patterned after Roman villas excavated from the ash of 79 C.E. One, in Saratoga Springs, New York, has an interior that was copied from the House of Pansa in Pompeii. Built in 1888 as a museum and tourist attraction, it later became a Masonic temple, then a Jewish synagogue. In recent years it has been occupied by local businesses.

The other building, in Malibu, California, near Los Angeles, is an almost exact replica of the Villa dei Papiri near Herculaneum. It was constructed in the 1970s by oil magnate

J. Paul Getty to house his large collection of Greek and Roman art. The building, itself a work of art, reproduces the original villa in its entirety, including interior courtyards and exterior gardens.

The Villa dei Papiri, among the largest private homes known from the Roman Empire, was situated outside the walls of Herculaneum, overlooking the Bay of Naples. It was excavated in 1754. Among the most important discoveries in the ruins was a small room containing more than 2,000 papyrus scrolls, badly burned but in some cases still readable. Most are treatises by a philosopher named Philodemus, who lived near the Sea of Galilee in the first century B.C.E. It is from those scrolls that the Villa dei Papiri gets its name, the original owner being unknown.

The lore of Vesuvius and Pompeii has been widely reflected in literature as well as in the decorative arts, clearly exemplifying our "vibrating string" metaphor. The awesome drama of Pompeii has been a magnet for writers. Sir Walter Scott is said to have muttered "The city of the dead! The city of the dead!" as he walked its streets. Charles Dickens, in *Pictures from Italy*, wrote, "the mountain is the genius of the scene. . . . we watch Vesuvius . . . as the doom and destiny of all this beautiful country, biding its terrible time." With a party of friends Dickens climbed to the summit of Vesuvius and, in the same book, described his impressions of the volcano's crater: "great sheets of fire are streaming forth: reddening the night with flame, blackening it with smoke, and spotting it with red-hot stones and cinders, that fly up into the air like feathers, and fall down like lead. What words can paint the gloom and grandeur of this scene!"[10]

The German writer Johann Wolfgang von Goethe, after climbing Vesuvius during a journey to Italy in 1787, wrote,

> we wandered about observing . . . features of this peak of hell which towers up in the middle of paradise.
>
> . . . A magnificent sunset and evening lent their delight to the return journey. However, I could feel how confusing such a tremendous contrast must be. The Terrible beside the

Beautiful, the Beautiful beside the Terrible, cancel one another out and produce a feeling of indifference. The Neapolitan would certainly be a different creature if he did not feel himself wedged between God and the Devil.[11]

Marie-Henri Beyle, a French writer better known as Stendhal, visited Pompeii in 1817 and later wrote, "Pompeii . . . is the most astounding, the most fascinating, the most entertaining spectacle I have ever encountered; no other sight on earth can furnish such *understanding* of antiquity."[12]

Among the earliest books about the catastrophe of 79 c.e. is *The Last Days of Pompeii,* a novel published to popular acclaim in 1834 by the English writer Edward Bulwer-Lytton. Though overly sentimental and melodramatic for modern tastes, it presents a fascinating glimpse of Pompeiian life in the first century and a vivid picture of what it must have been like when the earth shook, walls tumbled, and ash and lapilli rained down upon the city, turning day into night.

Bulwer-Lytton weaves a tale of one Glaucus, his betrothed Ione, and a blind flower-girl named Nydia, who is in love with Glaucus. In the climactic episode, as throngs of desperate people struggle through the darkness in an effort to escape from Pompeii, the blind Nydia, "accustomed, through a perpetual night, to thread the windings of the city," leads Glaucus and Ione to the seashore and safety aboard a boat. Then, her love unrequited, she quietly slips into the sea and perishes. Nydia was the subject of a play written in 1836 by an American, Henry Boker, and of a popular work in marble by the American sculptor Randolph Rogers in 1861. A similar novel in some ways, in content as well as title, is *Alive in the Last Days of Pompeii,* written in the 1970s by the British author Alan Lloyd.

The American writer Mark Twain visited Pompeii in 1867 and wrote,

There stands the long rows of solidly-built brick houses (roofless) just as they stood eighteen hundred years ago, hot with the flaming sun; and there lie their floors, clean-swept, and

not a bright fragment tarnished or wanting of the labored
mosaics that pictured them with the beasts, and birds, and
flowers which we copy in perishable carpets today; and
there are the Venuses, and Bacchuses, and Adonises, making
love and getting drunk in many-hued frescoes on the walls
of saloon and bed-chamber; and there are the narrow streets
and narrower sidewalks, paved with flags of good hard
lava, the one deeply rutted with the chariot-wheels, and the
other with the passing feet of the Pompeiians of by-gone
centuries.[13]

In 1909 another American writer, Henry James, published
an account of his travels in Italy called *Italian Hours,* in which
he wrote that the tourist in Pompeii, "with his feet on Roman
slabs, his hands on Roman stones, his eyes on the Roman void,
his consciousness really at last of some good to him, could
open himself as never before to the fond luxurious fallacy of a
close communion, a direct revelation."[14]

Books with Mount Vesuvius as centerpiece or background
continue to appear. In 1992, for example, another American
writer, Susan Sontag, published *The Volcano Lover,* a novel based
on the life of Sir William Hamilton. Sontag superbly captures
the essence of Hamilton's fascination with Vesuvius, his need
to write about it, his obsession with collecting artifacts from
Pompeii—and his friendship with Lord Nelson, Britain's great
naval hero, who for years openly maintained a scandalous
relationship with Lady Hamilton.

Mount Vesuvius even figures in an opera. *La Muette de Portici* (The mute woman of Portici), written in 1828 by the French
composer Daniel Auber, is notable for including the onstage
eruption of a mock-up of the volcano. The opera, sometimes
also known as *Masaniello,* depicts an uprising in 1647 by the
citizens of Naples, led by a fisherman named Masaniello,
against the Spaniards who then ruled them. The eruption,
coming at the end of the opera, symbolizes divine judgment.
A performance in Brussels in 1830 triggered student riots,
which in turn sparked the revolution that freed Belgium from
Dutch rule.

The destruction of Pompeii is a natural subject for motion pictures, and at least four—all named *The Last Days of Pompeii* or, in Italian, *Ultimi Giorni di Pompeii*—have been made over the years. The first was an Italian film made in 1897. Another Italian film was made in 1926, an American film in 1935, and still another Italian version in 1959.

Not surprisingly, the most eloquent thoughts about Vesuvius and Pompeii have been expressed by poets. In 1819 the British historian Thomas Babbington Macaulay wrote of Vesuvius,

> Saw ye how wild, how red, how broad a light
> Burst on the darkness of that mid-day night,
> As fierce Vesuvius scatter'd o'er the vale
> His drifted flames and sheets of burning hail,
> Shook hell's wan lightnings from his blazing cone,
> And gilded heaven with meteors not its own?[15]

And the English poet Percy Bysshe Shelley, after a visit to Pompeii in 1819, wrote,

> I stood within the city disinterred;
> And heard the autumnal leaves like light footfalls
> Of spirits passing through the streets; and heard
> The Mountain's slumberous voice at intervals
> Thrill through those roofless halls;
> The oracular thunder penetrating shook
> The listening soul in my suspended blood;
> I felt that Earth out of her deep heart spoke.[16]

Our fascination with Vesuvius, and with Pompeii and Herculaneum, is enduring. As Susan Sontag wrote in *The Volcano Lover*, "The mountain is an emblem of all the forms of wholesale death: the deluge, the great conflagration . . . , but also of survival, of human persistence. In this instance, nature run amok also makes culture, makes artifacts, by . . . petrifying history. In such disasters there is much to appreciate."[17]

Mount Vesuvius, and Pompeii and Herculaneum, continue to inspire writers and artists, stimulate scientists, and attract the merely curious. The volcano, and the cities it destroyed yet preserved, present a timeless juxtaposition of life, death, and resurrection.

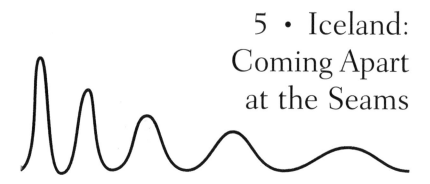

5 · Iceland: Coming Apart at the Seams

That long opening in the crust . . . was the Mid-Atlantic Rift.
It ran . . . along the longest range of mountains in the world,
twenty-eight thousand miles long, volcanically formed, and all
undersea. And the floor of the sea itself continued to split apart
at about the rate that a fingernail grows!

Joseph Hayes, *Island on Fire*

ICELAND IS A VOLCANIC ISLAND, about the size of the state of Virginia, that lies astride the Mid-Atlantic Ridge some 370 kilometers east of Greenland. It is among the world's most active volcanic areas. Since Norsemen settled there in the ninth century, there have been about 150 documented eruptions, many of them not from volcanic mountains but from fissures in the earth. The interior of the island, mostly uninhabited, is characterized by rugged terrain and several large ice sheets, or *jökulls*. Vatnajökull, in southeastern Iceland, has an area of 8,400 square kilometers and is a thousand meters thick in places. It is the third largest ice sheet on earth, after those in Antarctica and Greenland. Barren plateaus account for a little more than 50 percent of Iceland's area, lava fields 11 percent, and ice sheets 12 percent. With Iceland's frequent volcanic activity and widespread glaciers, it is not surprising that the country is known as "the Land of Fire and Ice."

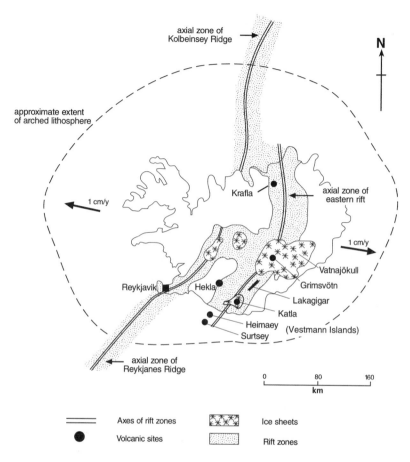

FIGURE 5-1. Tectonic setting of Iceland above its mantle plume, showing the principal rift zones and the volcanic sites discussed in this chapter. The North American and Eurasian plates are separating at a rate of about 2 centimeters a year (1 centimeter in either direction).

The North American and Eurasian tectonic plates are separating along the Mid-Atlantic Ridge. The eastern part of Iceland moves eastward with the Eurasian plate, and the western part moves westward with the North American plate. The country is literally being pulled apart at a rate of about 2 centimeters a year (Figure 5-1). But molten rock, or magma, continually erupts from fissures and volcanoes to fill the gap.

Iceland has the dubious distinction of being located not only upon a volcanic mid-ocean ridge but also above a hot spot—the surface manifestation of a plume of magma that rises from within the upper part of the earth's mantle. Seismic data suggest that the stem of the plume is about 300 kilometers wide and can be traced to a depth of about 400 kilometers. The upward force of the rising mass of molten rock has arched an area of the earth's lithosphere (the crust and solid upper mantle) with a diameter about twice that of the plume itself.

Some geologists believe the separating tectonic plates enabled the mantle plume to find its way upward, whereas others believe the rising plume caused the plates to separate. In any event, the greatest significance of Iceland for scientists is that it is the only place on earth where both a mantle plume and a mid-ocean ridge can be studied on land.

Immediately south of Iceland the plate boundary is located in a more or less continuous rift valley, which follows the axis of a broad segment of the Mid-Atlantic Ridge known as the Reykjanes Ridge. North of Iceland a similar feature, the Kolbeinsey Ridge, extends northward beneath the Greenland Sea. The structure of Iceland itself is more complex (see Figure 5-1). The Reykjanes rift zone enters the island from the southwest, near Reykjavik, and appears to end in central Iceland. The Kolbeinsey rift zone enters Iceland on the north coast and appears to end there. To the east is a third rift zone, which extends from northern Iceland to a volcanic archipelago known as the Vestmann Islands, or Vestmannaeyjar, off the south coast. Each of these broad rift valleys contains many swarms of fissures. East-west-trending faults, characterized by frequent earthquakes, extend from the apparent ends of the Reykjanes and Kolbeinsey rift zones to the rift zone in eastern Iceland.

The eastern rift valley contains Iceland's youngest volcanic rocks. Most were emplaced as lava flows that issued from long fissures in the valley. Eruption sites shifted intermittently as older fissures became clogged and younger ones opened alongside them to provide pathways for younger pulses of magma.

The oldest volcanic rocks, 16 to 22 million years in age, are found near the east and west coasts of Iceland, as would be expected on an island formed above separating tectonic plates. Flows below sea level, both eastward and westward, are still older. The oldest rocks created by volcanic activity above the more or less stationary mantle plume of Iceland are exposed along the west coasts of Scotland and Ireland and the east coast of Greenland and provide clear evidence of the opening of the North Atlantic Ocean. Those rocks are 40 to 50 million years old.

Volcanic rocks exposed on the other side of Greenland, along the west coast, are about 70 million years old and are also believed to have originated above the plume. Hence Greenland must have been located above the plume between 40 and 70 million years ago. Reconstruction of the movement of the North American plate over the mantle plume suggests that between 70 and 120 million years ago, the movement had a north-south component, and during that period first Ellesmere Island and then Baffin Island were over the plume (Figure 5-2).

Clearly it took a very long time for the plume to rise through the upper mantle. The thick continental lithosphere above the plume initially inhibited the dissipation of heat, resulting in high temperatures that created huge volumes of magma. When upward pressure finally ruptured the lithosphere, magma intruded into the many fractures and solidified there to form thin bodies of rock called *dikes;* and, forcing its way between layers of pre-existing rock, the magma formed massive, tabular bodies called *sills.* Wherever magma reached the surface, it produced enormous lava flows, which thickened the earth's crust and created the island now called Iceland.

Plume output waxed and waned. It appears to have been highest from 70 to 40 million years ago and again during the past 20 million years. Since the beginning of the ice ages about 2 million years ago, the plume has produced 400 to 500 cubic kilometers of lava, covering more than 12,000 square kilometers. Some 32 cubic kilometers of lava and about 10 cubic kilometers of fragmental volcanic debris are known to have surfaced

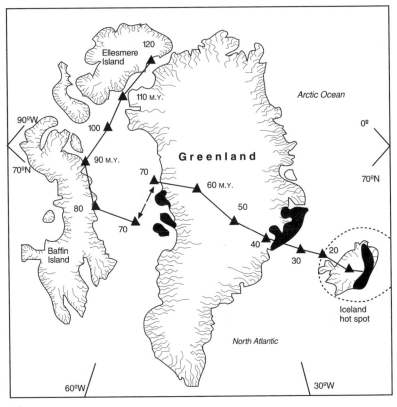

FIGURE 5-2. Plot showing approximate locations in Greenland, Baffin Island, and Ellesmere Island that once were above the Iceland hot spot. This diagram illustrates the changing directions in the movement of the North American tectonic plate as it has separated from the Eurasian plate—movement that has resulted in the opening of the North Atlantic Ocean. Adapted from Lawver and Mueller, "Iceland Hot Spot Track," 311.

from various volcanoes and fissures since Iceland's first settlers arrived from Norway in 874 c.e. Since that time, more than thirty volcanic centers have been active on or near the island. The volume of lava produced by individual eruptions has varied greatly, ranging from less than a cubic kilometer to as much as 20 cubic kilometers.

Volcanoes in Iceland usually have developed at the centers of large swarms of fissures. In contrast to fissure eruptions, which tend to pour forth their lava relatively quietly, Iceland's volcanoes tend to erupt explosively because their magma is less fluid. Two volcanoes in south-central Iceland illustrate this point. Katla, near the coast, has had ten eruptions with large volcanic explosivity indexes (VEIs), estimated to have been in class 4 or 5 ("large" to "very large"), since its first known eruption in 934 C.E. Hekla, about 56 kilometers northwest of Katla, has had nine such eruptions since its first recorded activity in 1104 C.E. Throughout Iceland there have been twenty-seven eruptions with VEIs of 4 to 5 since the country was settled—a rate of two or three per century. These events have greatly affected Iceland's population, and its culture.

An early volcanic catastrophe is reflected in a famous Icelandic poem, *Völuspá*, which appears in a thirteenth-century manuscript called the *Poetic Edda*. The poem is a retelling of Scandinavian mythology from the beginning of the world to its destruction and the death of the gods in a final disaster known as *Ragnarok*. The following stanza from *Völuspá* undoubtedly describes a major volcanic eruption:

> The sun begins to be dark; the continent falls fainting into
>   the Ocean;
> They disappear from the sky, the brilliant stars;
> The smoke eddies around the destroying fire of the
>   world;
> The gigantic flame plays against heaven itself.[1]

In all probability that description was inspired by a real cataclysm, memory of which subsequently faded into mythology. Scandinavian and German myths include many references to the destruction of the world, often involving fire and smoke that hides the stars, as well as prolonged frigid weather and the final battle of the gods. We know from historical eruptions, such as that of Tambora in Indonesia in 1815, that volcanic ash and aerosols in the atmosphere can filter out enough sunlight to significantly reduce the amount of heat reaching the earth and cause abnormally cold weather. These cold spells can persist for many months, even years. Irish author Padraic Colum vividly described such a scenario in his book *Orpheus: Myths of the World,* quoting an Icelandic myth about Ragnarok, the fate of the gods:

> Snow fell on the four quarters of the world; icy winds blew from every side; the sun and the moon were hidden by storms. . . . no spring came and no summer; no autumn brought harvest or fruit; winter grew into winter again.
>
> There was three years' winter. The first was called the Winter of Winds: storms blew, and snows drove down, and frosts were mighty. The children of men might hardly keep alive in that dread winter.
>
> The second winter was called the Winter of the Sword: those who were left alive amongst men robbed and slew for what was left to feed on; brother fell on brother and slew him; all over the world there were mighty battles.
>
> And the third winter was called the Winter of the Wolf. Then the ancient witch who lived in the Ironwood fed the Wolf Managarm on unburied men, and on the corpses of those who fell in battle. . . . The Heroes in Valhall would find their seats splashed with the blood that Managarm dashed from his jaws; this was a sign to the Gods that the time of the last battle was approaching.[2]

That final battle between gods and the giants, in Icelandic mythology, was to be followed by a time of peace and happiness. Odin, the supreme god, would create a new heaven and a new, fruitful earth where there would be no wickedness or misery.

The myth of Ragnarok may have grown out of a catastrophic eruption believed to have occurred in the ninth century and to have been followed by three years of icy summers, widespread famine, and bloody strife. The eruption may have come from Bardabunga, a volcano beneath the ice of Vatnajökull, which is thought to have had a major eruption around

900 C.E. Some geologists, however, believe it more likely that the event described was an eruption of Katla, which lies to the southwest, beneath a smaller ice sheet named Myrdalsjökull (see Figure 5-1). The resulting social and political disorder may well have led to the historic meeting of Icelandic chieftains in the year 930 at Thingvellir, about 40 kilometers northeast of present-day Reykjavik, that established the Althing, the world's first democratic parliament.

By 1000 C.E. the Althing was bitterly divided on the question of whether Icelanders should officially accept Christianity or continue to worship the ancient Norse gods. During one session, according to old records, when the debate was especially heated, a messenger arrived with the news that lava was erupting from a fissure near Reykjavik, threatening farmlands in the area. Followers of the old religion took the news as a sign that the gods were offended by the talk in favor of Christianity. But a Christian leader rose to his feet and, looking out over Thingvellir's desolate volcanic landscape, asked his colleagues what had angered the gods when the volcanic rock on which they stood had erupted in fire. It is said that his rhetorical question decided the vote in favor of Christianity.

The nineteenth-century British historian Thomas Carlyle wrote eloquently about the country of Iceland and the provenance of the Icelandic myths:

> In that strange island, Iceland—burst up, the geologists say, by fire from the bottom of the sea; a wild land of barrenness and lava; swallowed many months of every year in black tempests, yet with a wild gleaming beauty in summer-time; towering up there, stern and grim, in the North Ocean; with its snow-yokuls, roaring geysers, sulphur pools, and horrid volcanic chasms, like the waste chaotic battle-field of Frost and Fire—where, of all places we least looked for Literature or written memorials, the record of these things was written down. On the seaboard of this wild land is a rim of grassy country, where cattle can subsist, and men by means of them, and of what the sea yields; and it seems they were poetic men these, men who had deep thoughts in them, and uttered

musically their thoughts. Much would be lost had Iceland not been burst up from the sea, not been discovered by the Northmen![3]

The literature and music of Scandinavia and Germany have been influenced greatly by the mythology of Iceland. Probably the best known example is the four-part opera *Der Ring des Nibelungen* by the German composer Richard Wagner (1813–1883). The final part of Wagner's monumental work, *Götterdämmerung* (Twilight of the gods), is based in part on the *Poetic Edda*.

———————

Historically, the most active volcano in Iceland has been Hekla, which has erupted more than twenty times since Iceland was settled. Hekla has averaged about one eruption every forty years. Though the volcano rises only 1,497 meters from the sparsely vegetated highlands east of Reykjavik, the bleakness of its surroundings gives the rocky peak a menacing aspect. In the Middle Ages, volcanoes were considered gateways to the underworld. Descriptions of the black lava flows on the flanks of Hekla, its smoking summit, and the presence of flocks of ravens in the vicinity, taken to be souls of the dead, convinced many people that Hekla was indeed an entrance to hell.

In the twelfth century, a French cleric named Herbert of Clairvaux, in his *Liber Miraculorum* (Book of wonders), reported on Hekla as follows:

> This mountain, . . . all burning and belching flame, stands in a perpetual blaze, which spreads over the mountain and wastes it outside and inside. . . . That famous fire-kettle [Mount Etna] in Sicily, which is called the vent of Hell and to which, as has often been proved, the souls of the dying, condemned to burn, are daily dragged—men say that that is only like a little stove in comparison with this immense pit of Hell. . . . In our time it has been seen that it erupted so furiously that it destroyed most of the surrounding land. . . . Who is now so perverse and incredulous that he will not be-

lieve that eternal fire exists to make souls suffer, when with his own eyes he sees that fire of which we now speak?[4]

Herbert most likely heard about Hekla from a friend named Eskil, who was archbishop of Iceland. Almost four centuries later a German physician, Caspar Peucer, wrote,

> Out of the bottomless abyss of Hekla Fell, or rather out of Hell itself, rise miserable cries and loud wailings, so that these lamentations can be heard for many miles around. Coal-black ravens and vultures hover around this mountain and have their nests there. There is to be found the Gate of Hell, for people know from long experience that whenever great battles are fought or there is bloody carnage somewhere on the globe, then there can be heard in the mountain fearful howlings, weeping and gnashing of teeth.[5]

In 1675 a Frenchman named de la Martinière published a travel book in which he wrote that every now and then the devil would drag the souls of sinners out of the fires of Hekla and cool them on ice in the nearby ocean, presumably as a means of intensifying their torture when they were returned to the hellish underworld.

———

About 120 kilometers northeast of Hekla is another highly active volcano, Grimsvötn, which lies beneath the ice of Vatnajökull. The volcano has erupted twenty or more times since the fourteenth century, most recently in 1998. Its ice-hidden caldera has an area of 20 square kilometers and is 250 to 300 meters deep.

Heat from magma intruding beneath ice-covered volcanoes like Grimsvötn and Katla generates enormous volumes of meltwater. The heat of one cubic meter of magma can melt as much as 14 cubic meters of ice, creating 13 cubic meters of meltwater. The water accumulates in craters or calderas until it breaks through fissures in their flanks, or until eruptions discharge it, to form subglacial rivers that burst from beneath the surrounding ice sheets in enormous floods called *jökulhlaups* (glacier bursts), some of which can last for up to two weeks.

The volumes of some *jökulhlaups* are thought to have equaled the discharge rates of large rivers. Glacier bursts from Grimsvötn have carried masses of ice as well as sand, gravel, and large rock fragments down to the south coast of Iceland, where the debris has formed a vast outwash plain. Some of the debris, caught in blocks of ice, has floated out to sea and been found in drill cores taken from sediments on the ocean floor hundreds of kilometers from land.

In early October 1996 an eruption occurred along a long fissure beneath Vatnajökull about 15 kilometers northwest of Grimsvötn, where the ice is more than 500 meters thick. Subglacial topography directed meltwater from the eruption toward Grimsvötn's caldera, which rapidly filled. On November 5 an estimated 4 cubic kilometers of water, carrying enormous chunks of ice, burst from the caldera and surged southward. The onslaught, among the largest *jökulhlaups* ever seen in Iceland, washed over the uninhabited outwash plain like a juggernaut. Up to 5 meters deep in places, the rampaging water destroyed power lines, carried away reinforced-concrete bridges, and washed out at least 15 kilometers of roads. Fortunately there were no human settlements in its path.

Almost 200 years earlier, however, an eruption from fissures associated with Grimsvötn was perhaps as devastating as the event depicted in the thirteenth-century poem *Völuspá*. Starting in 1783 and continuing into 1784 there was a massive eruption from fissures that opened just southwest of Vatnajökull (Figure 5-3). The event created an appalling disaster. It began on June 8, 1783, after three weeks of earthquakes, and continued for eight months.

The fissure zone that opened in 1783 was almost 27 kilometers long and comprised ten individual segments, which appeared successively from southwest to northeast. Such progressive opening of individual fissures, followed by volcanism, appears to be characteristic of tectonic activity in Iceland's

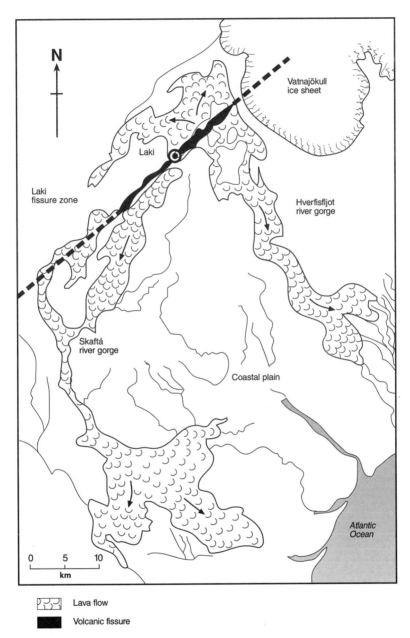

N

Vatnajökull
ice sheet

Laki

Laki
fissure zone

Hverfisfljot
river gorge

Skaftá
river gorge

Coastal plain

0   5   10
km

Atlantic
Ocean

⬚ Lava flow

■ Volcanic fissure

FIGURE 5-3. The Laki fissure zone and extent of lava flows of 1783 and 1784.

eastern rift zone. At the earth's surface, the fissure zone is characterized by at least 140 individual vents in the form of spatter cones (Figure 5-4) composed of volcanic ash and cinders as well as the glassy, rough-textured material known as scoria. The largest of the cones, as much as 90 meters high, must have been formed by fountains of fiery material that reached heights of more than a thousand meters.

Initially there were explosive eruptions of fragmental material along with large volumes of volcanic gases, which were propelled high into the atmosphere. The strongest eruptions had an estimated VEI of 4, and the volume of ejected material probably was about half a cubic kilometer. Atmospheric turbulence and prevailing winds dispersed dust and gases over much of Iceland and even as far away as Europe and North America. Volcanic ash and the heavier gases settled on Iceland's farms and caused much damage to crops.

In June 1783 enormous quantities of lava began pouring from twenty-two volcanic cones along a segment of a fissure zone crowned by a small, older volcano named Laki. The lava flows united to form a large stream that followed the winding gorge of the Skaftá River, which normally carries meltwater southward from Vatnajökull. Its waters turned to steam, and lava filled the gorge, 500 meters wide and 120 to 180 meters deep. From there the lava poured into a lake and filled it. In August a new batch of lava began flowing down the valley of a second river, the Hverfisfljót, which roughly parallels the Skaftá a few kilometers to the northeast (Figure 5-3). When the flows reached the coastal plain, they became 20 or more kilometers wide. People had settled those lowlands by 1783, and the spreading lava overran two churches and fourteen farms. Thirty more farms were damaged. The two rivers subsequently found new courses, their old valleys having been filled. Immense floods deluged the countryside, partly from melting of snow and ice and partly from the blocking of the river valleys.

A local pastor named Jón Steingrímsson witnessed the eruption and described it as follows:

FIGURE 5-4. The Lakagigar, or chain of volcanic spatter cones, above the Laki fissure. Photograph by Emanuela Baer, used with permission.

First the earth swelled up, with a chorus of howls, filled with an uproar that made it explode into pieces, tore it apart and eviscerated it like a rabid animal rips something to bits. Then flames and fires came out of the slightest hole in the lava. Big blocks of stone and sods of turf were thrown into

the air to an unspeakable height, from time to time accompanied by great bangs, flashes, jets of sand and light or dense
smoke. . . . The ground trembled frequently. Oh, what terror
it was to contemplate such signs of fury, such divine mani-
festations! . . . it was a suitable time to talk to God![6]

The eruption finally ended in February 1784, but gases
continued to rise from the cooling lava for several years. Some
565 square kilometers had been buried beneath 12 to 15 cubic
kilometers of lava. It was the greatest outpouring of lava on
earth in historic time. In addition to lava deposited on the
surface, an unknown volume of magma cooled within the fissures to form dikes or was injected between layers of adjacent
rock to form sills. The cumulative volume must have been
enormous because many fissures extend to depths of at least
10 kilometers.

The 1783–1784 event usually is referred to as the Laki
eruption, but in Iceland it is also called the *Skaftáreldar* (Skaftá
fires) because of fiery fountains that spouted from the row
of volcanic cones near Laki and the Skaftá River. The line of
eruption vents is known as the *Lakagigar* or, alternatively, the
Skaftáreldagigar (Figure 5-4).

During the eight months of the eruption the Laki fissures
released vast quantities of gases, some poisonous, into the
atmosphere. Atmospheric turbulence was caused both by the
eruption itself and by the temperature differential between
the hot volcanic material and the nearby ice of Vatnajökull.
The turbulence was so powerful that the gas-rich columns
reached heights of 12,000 to 13,000 meters and thus penetrated the lower stratosphere. Winds blowing primarily from
the northwest spread volcanic products southeastward toward
Europe. As a result, the amount of sunlight and therefore
heat was reduced, and the winter weather throughout Iceland, indeed all North Atlantic regions, was unseasonably cold
throughout 1783 and 1784.

The gases comprised mostly water vapor, but, by volume,
they also included 5 to 6 percent carbon dioxide, 3 percent

sulfur dioxide, and 1 percent hydrogen chloride and fluorine. It was those gases that created a true calamity in Iceland. Although sulfur dioxide accounted for only about 3 percent of the gases, there was an estimated 50 million tons of it in the air, and it produced a dry fog over Iceland, the North Atlantic Ocean, and parts of the adjacent continents. Moreover, much of the sulfur dioxide combined with atmospheric water vapor to form about 150 million tons of sulfuric-acid aerosols. Droplets of the acid rained down upon Iceland, increasing soil acidity and stunting the growth of the grasses that the country's livestock depended upon for feed. In addition, dense clouds of noxious, heavier-than-air gases such as carbon dioxide and fluorine created a bluish haze that filled the lowlands where most of the livestock grazed.

As a result of severely limited pasturage and fluorine-poisoned streams, half of Iceland's cattle and three-quarters of the nation's sheep and horses died during 1783 and 1784. Icelanders at that time subsisted mainly on ranching and fishing. The loss of such a major source of food led to what has come to be known as the "blue-haze famine." As Jón Steingrímsson reported, "the hairy sand-fall and sulphurous rain caused such unwholesomeness in the air and in the earth that the grass became yellow and pink and withered down to the roots. The animals that wandered around the fields got yellow-coloured feet with open wounds, and yellow dots were seen on the skin of newly-shorn sheep, which had died."[7]

Increased fishing could have saved many during the famine, but strong winds, unusually high waves, and longer-than-normal freezing of the harbors, all related to the volcanism, persisted well into 1785 and, most of the time, prevented fishermen from putting out to sea.

The Laki eruption occurred during a period when changing climatic conditions (the Little Ice Age) had already put a strain on Iceland's people. Lower seawater temperatures had thinned cod populations, and longer winters had reduced the amount of forage available for cattle. Volcanic ash and poison gases from the eruption abruptly worsened those conditions. More than a quarter of Iceland's 50,000 inhabitants died—a national catastrophe of colossal proportions.

———————

The eruption of 1783 had far-reaching effects because prevailing winds spread Laki's gases over much of the Northern Hemisphere. The dry, sulfurous fog slowly spread eastward and southeastward over Europe at an average rate of about 50 kilometers a day. There were several reports of damaged vegetation and crop failures at a most inopportune time in the growing season. A Dutch citizen named S. J. Brugmans, in a letter to Leiden University, wrote that in the town of Groningen, about 1,500 kilometers southeast of Laki, the fog was accompanied by a strong smell of sulfur, which caused headaches and difficulty in breathing. By June 25 the normally green countryside thereabouts looked desolate. Leaves had turned brown and had dropped from the trees.

Benjamin Franklin, who at that time represented the United States at the court of Louis XVI in France, noted the sulfurous haze that stung his eyes and hung over the country during the summer of 1783. The following winter he correctly correlated anomalous temperatures with the haze, and he suggested that they might be related to volcanic activity in Iceland. In May 1784, in a communication to the Literary and Philosophical Society of Manchester, England, Franklin wrote that the haze rendered the sun's rays so faint that "their summer effect in heating the earth was exceedingly diminished" and that

> Hence the surface was easily frozen.
>
> Hence the first snows remained on it unmelted, and received continual additions.
>
> Hence perhaps the winter of 1783–4 was more severe than any that had happened for many years.
>
> The cause of this universal fog is not yet ascertained. Whether it was adventitious to this earth, and merely a smoke proceeding from the consumption by fire of some of those great burning balls or globes which we happen to meet with in our rapid course around the sun, . . . *or whether it was the vast quantity of smoke, long continuing to issue during*

the summer from Hecla, in Iceland, . . . which smoke might be spread by various winds over the northern part of the world, is yet uncertain [italics added].[8]

Benjamin Franklin's suggestion constitutes the first documented recognition that volcanic activity might affect weather.

The winter of 1783–1784 was exceptionally cold in North America as well as in Europe. Average temperatures in the vicinity of Philadelphia were lower than normal during the fall of 1783, and between December 1783 and February 1784 the temperature plummeted to minus 4 degrees Celsius, a record. The Delaware River was frozen at Philadelphia from late December until mid-March. In New York, ice blocked the harbor for ten days. Baltimore's harbor remained frozen from January 2 to March 25, and indeed most of Chesapeake Bay was frozen. Even in Charleston, South Carolina, the harbor froze in February. Most remarkable was the freezing of the Mississippi River at New Orleans. After the ice broke up, ships encountered ice floes in the Gulf of Mexico as far as 100 kilometers south of that city.

In western Alaska, many Native Americans in villages along the Bering Strait, near present-day Nome, died of starvation in 1783, a year in which the arctic "summer" never came. The season for fishing, hunting, and collecting berries was cut short. The native peoples were unable to stock food for the next winter, or even to find enough to last them through what should have been their brief summer.

Atmospheric pollution from the Laki eruption caused mean temperatures to remain below normal in North America and Europe throughout 1784 and 1785. Additional evidence of volcanic pollution in the Northern Hemisphere has been obtained from ice cores in Greenland, in which there are layers that contain a significant concentration of sulfuric acid for the years 1783 and 1784.*

---

*Greenland ice cores have also indicated that an eruption in Iceland about 53 B.C.E. released almost twice as much sulfur dioxide as the Laki event. In all likelihood it, too, was followed by significant changes in weather patterns over much of North America and Europe.

Weather patterns created by the blue haze were enigmatic, however, for the summer of 1783 was warmer than usual in some parts of Europe. In 1788 Gilbert White, the vicar of Selbourne, a village in England, reported the following in a publication titled *The Natural History of Selbourne:*

> The summer of the year 1783 was an amazing and portentous one, and full of horrible phenomena; . . . the peculiar haze, or smoky fog, that prevailed for many weeks in this island . . . was a most extraordinary appearance, unlike anything known within the memory of man. . . . The sun, at noon, looked as blank as a clouded moon, and shed a rust-coloured ferruginous light on the ground. . . . All the time the heat was so intense that butchers' meat could hardly be eaten on the day after it was killed . . . and indeed there was reason for the most enlightened person to be apprehensive.[9]

The apparent contradiction between the observations of Franklin and White probably can be explained by differences in the effects of carbon dioxide gas in the lower atmosphere and acid aerosols at higher levels. The carbon dioxide that spread over Europe created a "greenhouse effect," trapping heat and causing warmer weather at first. But as the gas dispersed, the influence of the aerosol veils, which reflected heat from the sun, became increasingly important, resulting in cold weather.

The rift zone that includes Grimsvötn and the Laki fissures extends southwestward to the volcanic Vestmann Islands, a few kilometers off the southern coast of Iceland. The name Vestmann refers to early Celtic settlers, whom the Nordic inhabitants of Iceland called "west men." Celts from Viking settlements in the British Isles were among the first people to arrive in Iceland, along with Norsemen.

Indeed, Celtic monks from Ireland, seeking ascetic solitude for religious contemplation, sailed to Iceland as early as 800 C.E., more than seventy years before the first permanent settlers. In spirit, those monks followed the legendary St. Brendan, a sixth-century Irish cleric who, with a group of followers, was said to have sailed the North Atlantic in search of the "promised land of the saints" and to have discovered a mythical island now referred to as the Island of St. Brendan.

The following account of one of St. Brendan's voyages, from an ancient manuscript titled *Navigatio Sancti Brendani*, apparently describes a volcanic eruption at sea, most likely among the Vestmann Islands:

> They came within view of an island, which was very rugged and rocky, covered over with slag, without trees or herbage, but full of smiths' forges . . . they heard the noise of bellows blowing like thunder . . . Soon . . . one of the inhabitants came forth . . . ; he was all hairy and hideous, begrimed with fire and smoke. When he saw the servants of Christ near the island he withdrew into his forge, crying aloud: "Woe! Woe! Woe!" St. Brendan . . . said to his brethren: "Put on more sail and ply your oars more briskly so that we may get away from this island." Hearing this the savage man . . . rushed down to the shore, bearing in his hand a pair of tongs with a burning mass of the slag . . . , which he flung at once after the servants of Christ. . . . where it fell into the sea . . . a great smoke arose as if from a fiery furnace. . . . all the dwellers of the island crowded down to the shore, bearing . . . burning slag which they flung . . . after the servants of God; and then they returned to their forges, which they blew up into mighty flames, so that the whole island seemed one globe of fire and the sea on every side boiled up and foamed like a cauldron . . . and a noisome stench was perceptible at a great distance.[10]

Perhaps what St. Brendan saw was volcanism on an island that had just been born, as happened in November 1963, when fire and lava dramatically burst from the depths near the Vestmann archipelago. That eruption, with a VEI of 4, which volcanologists consider "large," created an island that was named Surtsey after Surtur, a giant who, in Norse mythology, came from the south carrying a flaming sword and brought fire to the northland.

The volcanic activity that created Surtsey was first noticed early in the morning on November 14. At about 7:00, a few kilometers southwest of Geirfuglasker, at that time the farthest south of the Vestmann Islands, the crew of a fishing boat noticed an offensive odor and felt irregular movements of the boat as the sea surged around it. The water was more than 120 meters deep there. After a few minutes, about a kilometer and a half to the southeast, they saw dark smoke rising from the sea. At first they thought a ship was afire, but then a column of ash, steam, and smoke burst forth, with flashes of fire. By 8:00 it had reached a height of 60 meters. The fishermen radioed the news to Icelandic authorities, and it was not long before aircraft carrying geologists and representatives of the news media were flying over the scene. By 11:00 A.M. the eruption column was almost 3,700 meters high. By 3:00 P.M. it was over 6 kilometers high and could be seen from Reykjavik, more than 110 kilometers away.

By November 15 a volcanic cone had risen from the depths to form a small island. It grew rapidly as eruptions continued, hurling volcanic bombs in all directions. An Icelandic geologist named Sigurdur Thorarinsson visited the scene aboard a coast guard vessel and wrote the following:

> The volcano was most vigorously active, the eruption column rushing continuously upwards, and when darkness fell it was a pillar of fire and the entire cone was aglow with bombs which rolled down the slopes into the white surf around the island. Flashes of lightning lit up the eruption cloud and peals of thunder cracked above our heads. The din from the thunderbolts, the rumble from the eruption cloud, and the bangs resulting from bombs crashing into the sea produced a most impressive symphony.[11]

On December 28 steam rose from the ocean about 2.5 kilometers northeast of Surtsey, about halfway between the new island and Geirfuglasker. That eruption ended early in January 1964 without the formation of another island. Meanwhile Surtsey continued to grow, and by mid-January the volcano had formed a sizable cone. Soon a glowing lava lake developed in the crater, and lava began flowing down the flanks of the cone. The future of Surtsey was now assured. Unconsolidated volcanic material is rapidly eroded by wave action during storms, but hardened lava can withstand thousands of years of wave battering. When the eruptions finally ceased late in 1965, Surtsey had reached an elevation of almost 170 meters and had an area of about 3 square kilometers. All together, as much as a cubic kilometer of lava and fragmental material may have been heaped upon the sea floor in that two-year period, creating the second largest island in the Vestmann archipelago.

Only one of the Vestmann Islands is inhabited. Named Heimaey (meaning "home island"), it has an area of more than 11 square kilometers and is the largest of the islands. In 1973 a fissure on Heimaey opened and produced fiery fountains of fragmental material followed by voluminous lava flows. Volcanic ash almost buried the island's town, which, like the archipelago itself, is called Vestmannaeyjar. Worse, the lava flows overran part of the town and threatened to close off its harbor. Destruction of the harbor would have been a national disaster, for Vestmannaeyjar is the most important fishing port in a country that depends upon fishing for much of its export trade.

In a heroic battle the people of Heimaey, aided by volunteers from all over Iceland and by equipment from the main island and other countries, pumped as much as 8 million tons of seawater onto the lava in an effort to cool the molten rock and stop it from flowing into the harbor. That was the only instance, on such a large scale, of human intervention in a volcanic eruption. An interesting novel about Heimaey's ordeal is *Island on Fire* by the American author Joseph Hayes. In one graphic passage, Hayes captures the essence of what the people had to contend with: "In the town [were] . . . fires beneath a chaotic pattern of arcing bombs, small blistering furnaces that had once been homes, and the figures of men running and turning puny crystal spouts of water onto the tiny infernoes dwarfed by . . . towers [of] flame and fiery cinders and black ash."[12]

Over the years the inhabitants of Heimaey had had little reason to be concerned about volcanism, as the one volcano on their island, Helgafell, had never erupted in historic time and was considered extinct. Geologists estimate that Helgafell was last active more than 5,000 years ago. Nevertheless, the people of Heimaey were concerned when Surtsey's volcanism moved northeastward toward Geirfuglasker in 1963—and, as it turned out, it was only ten years later that new fissures allowed magma to erupt on Heimaey. It was early in the morning of January 23, 1973, that the fissures ripped open to a length of a kilometer and a half, traversing the eastern part of the island from shore to shore (see Figure 5-5). The eruption was only about 200 meters east of the town of Vestmannaeyjar, home to 5,300 people. The northward progression of volcanism from Surtsey to Geirfuglasker to Heimaey was not unlike a re-enactment of the ancient myth of Surtur, the giant with the flaming sword who came from the south, bringing fire.

As usual, the eruption was preceded by low-magnitude earthquakes caused by the opening of fissures at depth. A swarm of tremors occurred on January 22, and there were additional shocks four hours before the beginning of the eruption. The strongest were felt about 1:45 A.M., fifteen minutes before molten magma reached the surface. An Icelandic author, Arni Gunnarsson, wrote,

> The ground began to quiver slightly. . . . Then the surface of the land appeared to swell up, and soon it began to crack. It was as if a sharp knife were being drawn over the flesh of the earth and blood were beginning to spurt. Only, it was not blood; it was fire and embers. In a matter of minutes, nearly a mile-long fissure opened. . . . Lava immediately began to well out of it, and glowing cinders squirted high into the air.[13]

Studies of the depths of the focal centers where the earthquakes originated suggest that the magma had surged upward some

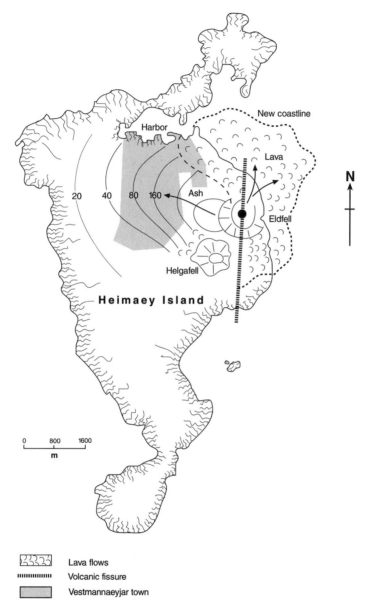

N

FIGURE 5-5. Heimaey Island, showing where fissures opened in 1973. Lava from vents along the fissure zone threatened to cut off the island's harbor from the sea and added almost 2.5 square kilometers to the area of Heimaey. Contour lines show approximate thickness, in centimeters, of fragmental volcanic material deposited on the town of Vestmannaeyjar.

10 kilometers in less than twenty-four hours. The shaking was strong enough to sever undersea pipelines and cables from the mainland that supplied Vestmannaeyjar with fresh water, electricity, and telephone service.

During the first three days spectacular curtains of fiery lava shot high into the air from several vents, but the eruption soon became concentrated toward the middle of the fissure, about 800 meters northeast of Helgafell. Large quantities of material were thrown into the air, ranging from fine ash to sizable bombs of molten lava. Within two days a cinder cone had formed and had risen to more than 100 meters above sea level. It was named Eldfell, meaning "fire mountain." The maximum VEI of the eruptions that created Eldfell was about 4, rated "large" by volcanologists. The volcano ultimately reached a height of 224 meters and today is essentially a twin of Helgafell, which itself remained inactive throughout the events of 1973.

Within a few hours of the start of the eruption, fishing boats and aircraft ferried most of Heimaey's inhabitants to safety on the mainland. Two or three hundred volunteers remained behind—and indeed, Icelanders from all over the country came to Heimaey—to help save the town of Vestmannaeyjar and its vital economy. The eruption continued for more than five months, producing an estimated 250 million cubic meters of fragmental material and the rough-textured lava known by its Hawaiian name, aa. Volcanic ash covered the town like black snow, as much as 5 meters deep in places. Streets were filled with ash, and, as in ancient Pompeii, houses were buried, or nearly so. Some could be identified only as hillocks in the ash. A winter snowstorm created a bizarre black-and-white landscape.

Red-hot lava bombs continually rained down upon Vestmannaeyjar, plunging through the roofs of buildings and keeping firefighters busy. As in wartime, people out of doors kept an eye peeled for incoming missiles. They learned not to run from them, but to simply watch until they were sure of their trajectory, then step aside if necessary.

Molten lava poured from the fissure at a rate of perhaps 40 to 50 cubic meters per second, flowing eastward and northward into offshore waters during January and February and eventually adding two and a half square kilometers to the area of Heimaey. In March another lava flow from Eldfell surged to the northwest, invading the eastern sector of Vestmannaeyjar. Slowly, inexorably, a wall of blocky lava some 20 meters high devoured a great many houses and one of the town's three fish-processing plants. The threat to Vestmannaeyjar's harbor, however, was of more desperate importance. The lava threatened to close the harbor's entrance.

The town of Vestmannaeyjar has the only good harbor along the entire southern coast of Iceland. Heimaey, moreover, is located almost in the center of Iceland's most productive fishing grounds and over the years has produced 8 to 10 percent of Iceland's export income. Abandoning the island to the caprices of Eldfell was simply not an option for the Icelanders.

Officials discussed a number of ideas for diverting the lava from the town. One of them was to bomb the volcano. From the NATO base at Keflavík, only about 130 kilometers from Heimaey, U.S. Air Force bombers could possibly blast away the east side of the mountain. Thus the lava, it was thought, would flow harmlessly into the ocean. Or U.S. Navy vessels could shell the volcano. But both schemes were rejected because Eldfell was too close to the town, and the results would be unpredictable.

An alternative scheme was to bomb the front of the underwater lava flow that directly threatened Vestmannaeyjar's harbor. Presumably bombing would break the insulating crust at the top of the lava and allow the red-hot lava inside to cool more rapidly. The government of Iceland and the U.S. Navy had a plan in place, but the day before the plan was to have been implemented, scientists pointed out that the results of abruptly mixing seawater and molten lava might well be explosive, and indeed catastrophic. So that plan, too, was abandoned. As explained by S. A. Colgate of the New Mexico Institute of Mining and Technology and Thorbjörn Sigurgeirsson of the

University of Iceland, "we realized the awesome possibility that once mixing was initiated it might be self-sustaining in that the high pressure steam produced might cause further mixing until all the lava had exchanged its heat with the water above it. The energy released might have come to between 2 and 4 megatons."[14]

The horror of such an explosion can perhaps best be illustrated by comparison with the atomic bomb that destroyed the Japanese city of Hiroshima in 1945. That bomb generated an amount of energy equivalent to 15,000 tons of TNT, or 15 kilotons. A 2-megaton explosion would be more than 130 times as large.

In early February, someone suggested spraying water on the lava to cool it and slow, or perhaps even stop, its advance. Fire engines were brought to the advancing lava front and, as an anxious nation watched on television, firefighters played hoses on the slowly moving molten mass. Their puny effort had little effect and became the subject of dark humor. A national joke had it as *pissa a hraunid, a hraunid* meaning "on the lava."

The water *did* have a cooling effect, however, and some of the lava solidified more rapidly at the very front of the flow. The problem was that the enormous molten mass behind the front just kept coming, surging over what little lava had hardened. That first effort was promising, but much more water was needed if the town and its harbor were to be saved.

Early in March a dredging ship, the *Sandey,* was brought into Vestmannaeyjar harbor from Reykjavik. Its powerful pumps began pouring seawater onto the lava at the rate of about 20,000 liters a minute. Giant pumps were flown in from the United States as well and were installed along the shore and on barges in the harbor. To get the water where it was needed, far back from the front of the flow, bulldozers hauled lengths of pipe over the crust of the advancing lava.

Only about half a meter of solid but still hot rock on top of the moving flow separated workers and bulldozers from molten lava with temperatures of almost 1,000 degrees Celsius.

The soles of workers' boots burned, and the tracks of the machines became so hot that the steel turned blue, but the desperate work went on. The evaporating water formed great clouds of steam, which added to the difficulty of the dangerous work on the lava flow.

By early April a thousand liters of seawater were being pumped onto the advancing lava every second. The water seeped into fractures in the crust, cooling the lava at the surface to about 100 degrees Celsius, forming the volcanic rock called basalt. Basalt is fluid at temperatures of 1,000 to 1,200 degrees, but essentially it ceases to flow at temperatures below 800 degrees. Gradually the advancing lava began moving more slowly.

In mid-February part of the north flank of Eldfell collapsed. Like icebergs calving from a glacier, chunks of solidified debris broke loose and slowly floated northward on the lava flow. The largest piece resembled a small mountain with a sharp peak and a base that covered more than 3.5 hectares. It was estimated to have weighed about 2 million tons. Someone dubbed it *Flakkarinn*, "the wanderer." As Flakkarinn ponderously floated northward it pushed up waves in the molten lava before it, and fresh lava filled the trough that developed behind. In two weeks the floating minimountain had covered three-quarters of a kilometer. People are said to have climbed onto Flakkarinn and ridden it.

Flakkarinn was headed directly toward the entrance to Vestmannaeyjar's harbor. Should it get that far, all efforts to save the harbor would be in vain. Lava waves ahead of Flakkarinn would overwhelm the barriers that had been pushed up, and Flakkarinn itself would surely block the harbor. All available pumps were put to work and poured about a hundred million liters of water onto the lava southeast of the harbor entrance. The water cooled and solidified enough lava to create a basalt barricade strong enough, it was hoped, to stop Flakkarinn. When the collision came, it was not unlike the proverbial encounter between an irresistible force and an immovable object. Flakkarinn, with elephantine slowness, spun halfway around and finally broke into pieces. The basalt dam held. Vestmannaeyjar's harbor was saved.

Lava production began to decrease after the first week of February, and by summer the eruption had stopped altogether. Some volcanic activity may have continued along a submarine segment of the fissure, however. On May 26 the crew of a fishing boat noticed short-lived submarine activity about 6 kilometers northeast of Heimaey, only 3 kilometers from the coast of Iceland. The northeastward progression of volcanic activity along a fissure was similar to that following the eruption of Surtsey and the Laki event of 1783 and 1784, again suggesting a factual basis for the ancient myth of Surtur's bringing fire from the south.

Where lava from Eldfell flowed into the ocean, it increased the size of Heimaey by almost 20 percent. Today there are basalt cliffs 30 meters high where, before 1973, fishermen plied their trade. Waves sometimes break over the cliffs during storms, and already there are beaches of wave-rounded rocks and pebbles at the base of the cliffs in places.

Vestmannaeyjar's harbor, which so many people worked so valiantly to save, is today more secure from storms than ever. Before the eruption the harbor entrance was more than three-quarters of a kilometer wide and was exposed to east winds. Now, thanks to the lava flow, it is only about 150 meters wide and better sheltered.

Within ten years of the eruption, Vestmannaeyjar was once again the leading fishing port in Iceland. The efforts of its citizens to control a volcanic eruption and save their town established procedures that one day may help save other towns in other countries. And Eldfell and its lava flows have become an important natural laboratory for geologists, a major tourist attraction, and a poignant testimonial to an indomitable people.

———

The cataclysm of *Völuspá*, the Laki catastrophe, the birth of Surtsey, the heroic defense of Vestmannaeyjar—all are part of the saga of Iceland, the island country that straddles the Mid-Atlantic Rift. Geologically, half of Iceland is part of the Eurasian tectonic plate and half is part of the North American plate. As those plates inexorably separate, the island is slowly being torn apart. Simultaneously magma wells up from the depths of the great rift. As the new magma finds its way to the surface through fissures and volcanoes, it keeps welding the halves together, continually regenerating the Land of Fire and Ice.

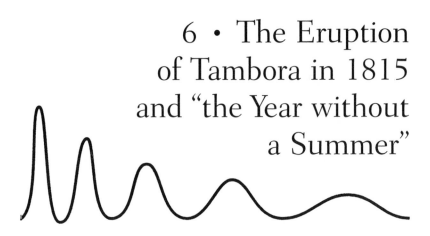

6 · The Eruption of Tambora in 1815 and "the Year without a Summer"

The bright sun was extinguish'd, and the stars
Did wander darkling in the eternal space,
Rayless, and pathless, and the icy earth
Swung blind and blackening in the moonless air;
Morn came and went—and came, and brought no day,
And men forgot their passions in the dread
Of this their desolation.

Lord Byron, "Darkness"

IN NORTH AMERICA IN 1815, the United States was a young country that extended no farther west than Ohio, Kentucky, Tennessee, and Louisiana; and eastern Canada was a British colony. Both the United States and Canada were agricultural countries with few cities. Most people lived on farms, and their livelihood was intimately linked to the weather.

On the other side of the world that April, on the island of Sumbawa in what was then the Dutch East Indies, now Indonesia, a volcano named Tambora exploded in the greatest eruption known to history. It obliterated entire populations, destroyed forests, and covered rice fields with volcanic ash. On Sumbawa and neighboring islands, possibly more than 70,000 people died—killed outright by the effects of the eruption or, in the months that followed, victims of starvation and disease.

The volcanic blasts sent great quantities of dust and gases high into the stratosphere, where for several years they circled the globe and reduced the amount of solar radiation that could reach the earth's surface. Worldwide temperatures dropped, and patterns of rainfall changed dramatically. Dry areas became wet, and wet areas, dry. Crops failed in many countries the following year, 1816. In Europe, famine and social unrest plagued countries still reeling from the devastation wrought by the Napoleonic Wars, which ended in June 1815 with the Battle of Waterloo.

In North America 1816 became known as "the year without a summer." The changed weather patterns were disastrous to farmers, especially in the Northeast. There were snowstorms and killing frosts in June, July, and August. People and livestock died from starvation. New Englanders, with wry Yankee humor, called it "eighteen hundred and froze to death." Crop failures accelerated a major population shift already underway as people migrated from New England to Ohio and the territories farther west.

But worldwide communications were still primitive. There was no way of relating the weather in Europe and North America to the eruption of far-off Tambora.

———

Today's Republic of Indonesia, the former Dutch East Indies, comprises more than 13,000 islands, about half of which are inhabited. The island chain extends more than 5,000 kilometers along the equator between Australia and the mainland of Southeast Asia. Known as "the string of emeralds," the islands include Sumatra, Java, Kalimantan (formerly Borneo), Sulawesi (formerly Celebes), and Irian Jaya (formerly Dutch New Guinea). There are more than seventy-six historically active volcanoes in Indonesia, mostly on Sumatra, Java, and the Lesser Sunda Islands, which include Bali, Lombok, and Sumbawa, the site of Tambora.

The fertile volcanic soils, abundant rainfall, and tropical climate enable most areas to produce two or three crops a year, not only of rice but of sugar cane and a variety of vegetables. In addition, large areas are devoted to the cultivation of coffee, tea, tobacco, and spices. Most of the land was originally covered with tropical rain forest, much of which has been and continues to be exploited for lumber.

The islands of Indonesia were home to some of the earliest known humanoids, once called *Pithecanthropus erectus* or "Java Man" but now classified as *Homo erectus.* They no doubt migrated from the mainland of Asia during the ice ages, when the shallow Sunda Sea dried up. Human migration continued, but when sea levels rose after the last ice age, populations on the various islands became isolated. Widely different languages and cultures developed, ranging from the Dajak headhunters of Kalimantan to the refined artistic communities of Bali.

Eventually European demand for spices led Portuguese, English, and Dutch explorers to the Indies, or Spice Islands, as they were called. As a consequence, the native peoples gradually came to be subjugated by Europeans—especially, by the end of the eighteenth century, the Dutch. In 1811, during the Napoleonic Wars, the British occupied Java. They replaced the governor, who had been appointed by Napoleon, with an Englishman named Thomas Stamford Raffles and gave him the title of lieutenant governor. Raffles, who later went on to found Singapore and earn a knighthood, governed the islands until 1816, when their administration was returned to the Dutch. Thus Raffles was in office when Tambora erupted in 1815.

Native priests on eastern Java interpreted volcanic phenomena associated with the eruption as signs that the gods would soon free the islands from European rule. Freedom, however, would have to wait another 130 years, until after World War II. Japanese troops conquered the islands early in that war, and political instability following their surrender in 1945 led to a nationalist revolution. Independence was proclaimed then, but it was not until 1949 that the Dutch formally transferred sovereignty to the Indonesians.

———

Previous eruptions of Tambora created most of the Sanggar peninsula on the north shore of Sumbawa, one island in the chain that makes up the Indonesian volcanic arc. That long,

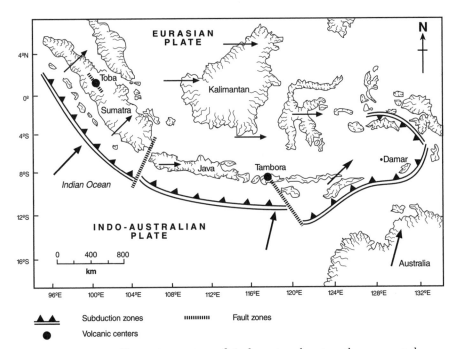

FIGURE 6-1. Tectonic setting of Indonesia, showing the present-day northeasterly drift and subduction of the Indo-Australian plate beneath the eastward-moving Eurasian plate and the resulting volcanic centers of Tambora and Toba.

southward-convex island arc extends from Sumatra to the island of Damar, where it curves sharply north and continues as a string of submarine volcanoes (Figure 6-1). The arc developed above the zone where the giant tectonic plate of Eurasia and the smaller Indo-Australian plate collide. Satellite technology tells us that the Eurasian plate is moving eastward about 2.5 centimeters each year while the Australian plate is moving northeastward almost 8 centimeters a year. The easterly component of Australia's motion is thus faster than that of Eurasia. As a consequence, part of the Australian plate is gradually slipping beneath the Indonesian margin of the Eurasian plate. It is this process of subduction that causes volcanism in Indonesia as very hot fluids, mostly water, are driven out of the subducted slab. As explained in Chapter 1, the hot fluids move upward and, through chemical interactions, decrease melting temperatures in the overlying wedge of mantle rock, thus forming the molten rock known as magma.

Tambora is some 340 kilometers north of the Java Trench, along which the subduction is occurring, and 160 to 190 kilometers above the down-sliding Australian plate. For magma to rise from the underlying mantle into the earth's solid crust, there must be faults in the crust. Tambora developed above the Sumba fault zone, which trends northwest-southeast, passing west of the island of Sumba and offsetting the Java Trench (Figure 6-1). Such faults, which cut across prevailing tectonic trends, are commonly referred to as cross faults.

The Sumba fault zone developed because of differences in the makeup of the Australian plate. West of Sumbawa, beneath the islands of Lombok, Bali, and Java, the plate's lithosphere—the crust and solid upper part of the mantle—is of oceanic origin. It is relatively thin and dense. East of Sumbawa, however, beneath Sumba, Flores, and Timor, the Australian lithosphere is of continental origin—thicker and less dense, therefore more buoyant. Those differences have caused differential motion and the development of a fracture zone between the oceanic and continental parts of the plate during its northeastward subduction. Such cross faults commonly propagate into the overlying plate margin and provide ideal pathways for rising magma.

The age of Tambora is unclear. The oldest lava exposed on the volcano is about 50,000 years old. The mountain is no doubt much older than that, however. Over the eons, a volcano developed that probably rose as high as 4,000 meters above sea level before the 1815 eruption. It was among the highest volcanoes in the East Indies and a noted landmark for ships at sea.

The youngest volcanic rocks on Tambora are massive layers of ash and coarser debris from the 1815 eruption. They overlie volcanic rocks that are believed to be about 5,000 years old. This age difference implies a long period of quiescence preceding the titanic explosion of 1815. That eruption reduced the height of the mountain to its present elevation of 2,853 meters and left a huge caldron-shaped depression, or caldera, 1,200 meters deep. (Some reports suggest that before the 1815 eruption, a caldera more than 43,000 years old already existed on Tambora.)

The volcano's reawakening apparently began sometime in 1812 with earthquakes and mild eruptions of steam in the crater, followed by intermittent explosions and dark clouds of volcanic ash. Such minor eruptions indicate that slowly rising magma had come into contact with water that had penetrated deep into the ground. The resulting build-up of steam pressure caused the periodic outbursts. Ash from those eruptions accumulated on the higher flanks of the volcano but was only a few centimeters thick.

The first significant eruption occurred during the evening of April 5, 1815. The distribution, composition, and grain size of deposits from that event indicate a violent eruption that produced a column of ash and smoke that may have been as much as 25 kilometers high. The explosions were heard in Batavia (now Jakarta), almost 1,300 kilometers to the west on the island of Java, and on the small island of Ternate, 1,400 kilometers to the northeast.

Soldiers garrisoned on eastern Java were so alarmed by the detonations that they went looking for revolutionaries, and in Batavia Lieutenant Governor Raffles sent boats into the Java Sea to search for a possible ship in distress. At Makassar, on the large island of Celebes (now Sulawesi), which lies between Java and Ternate, an armed vessel was dispatched to look for pirates.

That initial eruption was followed by five days of relatively low-level activity, producing minor ash falls. On the evening of April 10, however, a series of prodigious explosions began with a column of smoke, ash, and pumice that reportedly shot more than 40 kilometers into the air. Large eruptions followed for several days. The most powerful probably had a volcanic explosivity index (VEI) of 7, which volcanologists term "colossal."

After each outburst fiery clouds of gas, ash, and larger particles sped down the flanks of the mountain. Known as pyroclastic flows, they spread over much of the peninsula and entered the surrounding seas. They covered the sea floor, destroying all aquatic life within several tens of kilometers of the coast. Reacting with cold seawater, the hot material exploded into secondary eruptions that sent large quantities of fine ash into the atmosphere. The volume of ash produced by these secondary eruptions is thought to have been about ten times greater than the amount of ash generated by the original eruption. The total volume of ash produced by eruptive clouds and pyroclastic flows, then, may have been the equivalent of 50 cubic kilometers or more of dense rock.

Such an enormous volume indicates that Tambora's eruption probably was an order of magnitude greater than that of Krakatau in 1883 and two orders of magnitude greater than the eruption of Mount St. Helens in 1980. The actual volume, however, is problematical. The volcano's magma chamber is thought to have been shallow, less than 5 kilometers deep. Its roof collapsed during the eruption and left a caldera about 6 kilometers wide and 1,200 meters deep. It has been estimated that Tambora lost about a quarter of its original 4,000-meter height. If so, the corresponding missing volume must be about 34 cubic kilometers. However, the total mass of ejected material, as indicated above, corresponds to about 50 cubic kilometers—substantially more than the present volume of the caldera. The discrepancy might be explained by an injection of fresh molten rock into the magma chamber after the eruptions that began on April 10, raising the floor of the original caldera.

The eruptions that began on April 10 were heard as far away as western Sumatra, more than 2,500 kilometers west of Tambora. The earth trembled, and atmospheric shock waves shook houses 800 kilometers away in eastern Java. Ash and dust continued falling intermittently, with gradually lessening intensity, until April 17.

In Sanggar, a small town some 40 kilometers east of the eruption center, the local raja later described three columns of fire that rose high into the sky, creating a veritable firestorm over the volcano. He reported that moments later a sea of fiery lava (probably pyroclastic flows) spread over the entire mountain. After a terror-filled hour, according to the raja's account, dark clouds of volcanic ash descended upon the town, and stones, some of fist size, rained from the sky. The ashfall increased rapidly and was accompanied by turbulent hurricane-force winds, which ripped bamboo houses from their foundations, uprooted trees, and carried people, cattle, and entire coastal villages into the sea. Pyroclastic flows that slammed into the sea created tsunamis that ran up to heights of almost 5 meters. The waves ripped fishing boats from their moorings and carried windblown debris back inland.

People in the village of Bima, 65 kilometers east of Tambora, were shaken by loud, almost continuous detonations during the night of April 10, a "night" that lasted almost four days while the area remained shrouded in a dense ash cloud. The weight of fallen ash collapsed the roofs of most houses in the village, and, as in Sanggar, tsunamis flooded the coast and ravaged the lowlands. Villages were abandoned throughout Sumbawa, as most of the houses had been destroyed. An investigating party dispatched by Raffles found countless corpses of people and animals lying on the ground and floating in the sea.

In September 1815 the lieutenant governor presented a report on the eruption to the Natural History Society in Batavia. The British scientist Charles Lyell (1797–1875), among the founders of modern geology, incorporated some of Raffles's information in describing the effects of the eruption in his epochal *Principles of Geology,* as follows:

> In April, 1815, one of the most frightful eruptions recorded in history occurred in the province of Tomboro [Tambora], in the island of Sumbawa. . . . Out of a population of 12,000, in the province of Tomboro, only 26 individuals survived. Violent whirlwinds carried up men, horses, cattle, and whatever else came within their influence, into the air; tore up

the largest trees by the roots, and covered the whole sea with floating timber. Great tracts of land were covered by lava [pyroclastic flows], several streams of which, issuing from the crater of the Tomboro Mountain, reached the sea. . . . The floating cinders to the westward of Sumbawa formed, on April 12th, a mass 2 feet thick, and several miles in extent, through which ships with difficulty forced their way.

The darkness occasioned in the daytime by the ashes in Java was so profound, that nothing equal to it was ever witnessed in the darkest night.[1]

In 1855 a missionary named Heinrich Zollinger estimated that on Sumbawa as many as 10,000 people had been killed by the eruption.[2] On Sumbawa and the neighboring islands of Lombok and Bali, acidic ash poisoned rice fields and clogged their intricate systems of irrigation, destroying the crops on which people and livestock depended for food. Tambora deposited as much as 100 centimeters of ash on Sumbawa, 60 centimeters on Lombok, and 30 centimeters on Bali. The resulting famine and associated diseases claimed another 38,000 lives on Sumbawa, according to Zollinger, and 36,000 survivors left that island for Java. On Lombok as many as 20,000 people died, and perhaps 100,000 migrated to Java. Because of Java's already high population density, these mass migrations led to conflicts when the refugees landed on Java's eastern shores.

Ash thickness decreased with increasing distance from Tambora, but small amounts reached southern Sumatra, southern Kalimantan and Sulawesi, and western Timor. The ash covered an area of more than 500,000 square kilometers (Figure 6-2). The ash was so thick in the atmosphere that within about 300 kilometers of Tambora there was almost total darkness for three days.

Overlying the ash on Sumbawa is a sequence of partially consolidated pyroclastic-flow deposits, which locally reach a thickness of 30 meters. Those deposits are exposed in cliffs along the coast of the Sanggar peninsula.

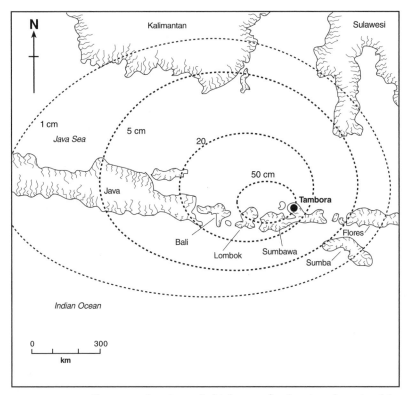

FIGURE 6-2. Extent and estimated thickness of volcanic ash emitted by Tambora in 1815. Adapted from Zollinger, *Besteigung des Vulkanes Tambora,* and Self et al., "Volcanological Study of the Great Tambora Eruption of 1815," 661.

Pumice ejected from Tambora floated on coastal waters, and rafts of the frothy material extended far into the Java Sea. Dead animals, broken trees, and other debris were imbedded in the rafts, some of which were several kilometers long and as much as a meter thick. Ships encountered them as many as four years after the eruption.

Along with the ash and pumice, the volcano emitted great quantities of steam and sulfur dioxide gas. Particles of sulfurous salts formed in the eruption column and adhered to bits of volcanic debris. The debris returned to the earth during rainstorms, poisoning the soils and groundwater of Sumbawa and nearby islands for years. The sulfur was responsible for deadly outbreaks of diarrhea in both people and animals.

On Sumbawa virtually all vegetation was destroyed or severely damaged in the catastrophe. Large quantities of water normally evaporate from the pores of plant tissue through the process of transpiration. After the forests were defoliated and the ground covered with volcanic ash, the volume of water evaporated into the atmosphere over Sumbawa decreased rapidly, which resulted in fewer clouds and greatly reduced rainfall. Although new soils can form relatively quickly on tropical islands, the drought and chemical changes in Sumbawa's soil delayed the recovery of vegetation for decades.

The eruption caused major weather changes around the world. Its effects upset the regular pattern of the summer monsoon winds in India, which bring much-needed rain to the subcontinent. Instead of the normal rainfall, there was drought in several areas during the summer of 1816. Then September brought almost incessant rain and severe flooding, notably in what is now Bangladesh. Grain became scarce and famine ensued, especially in northwestern India (present-day Pakistan). In China that summer there were major floods in the valleys of the Yangtze and Yellow rivers (now called Chang Jiang and Huang He).

Moreover, there was an outbreak of cholera in India's lower Ganges valley in 1816. The endemic disease rapidly became epidemic, in all probability because failed harvests and the ensuing famine had weakened the population. Cholera soon spread to Afghanistan and Nepal, no doubt facilitated by British military operations in those areas, and Moslem pilgrimages carried it farther west, to Mecca and Medina in Arabia. The disease advanced slowly but steadily northward. By 1823 the cholera appears to have reached the shores of the Caspian Sea. There was an outbreak in Moscow in 1830.

The following year in Egypt, Cairo lost 12 percent of its population to cholera. Russian military campaigns spread the disease westward into Poland, and from there it reached Hungary and eventually France, where hundreds of thousands of people perished. The country's prime minister succumbed in May 1832. Early that summer in North America, immigrants brought cholera to Montreal and New York City. By July there were as many as 100 deaths a day in New York alone.

It is not certain, of course, that the eruption of Tambora was responsible for both waves of cholera. But even if only the first epidemic, which raged from 1817 to 1823, was related to the weather changes and famine that followed the eruption, there can be little doubt that Tambora claimed several hundred thousand lives.

———————

High in the stratosphere after Tambora's eruption, sulfur dioxide molecules combined with water vapor to yield sulfuric acid aerosols (Figure 6-3). Prevailing winds carried the aerosols around the world. They formed veils that reflected a significant amount of sunlight, preventing its warmth from reaching the earth's surface. And because the veils remained above the clouds and were not washed out of the atmosphere by rain, they remained in place for years and created a long-term cooling trend.

Once the acidic aerosol droplets began to sink lower in the atmosphere, moreover, they provided nuclei for condensation and hence increased cloud formation. Acidic droplets are smaller than the droplets in normal clouds, and many small droplets are much more reflective than fewer large droplets. Thus the clouds themselves became more reflective, which added to the cooling trend.

Weather data for the early nineteenth century indicate a two- to three-year period of weather extremes following the eruption. Throughout 1816, average surface temperatures in the Northern Hemisphere were as much as 10 degrees Celsius lower than normal. A global cooling trend had been in progress for several years before the eruption of Tambora. The Tambora event accelerated that trend.

Throughout Europe, the summers of 1816 and 1817 were cold and wet. Snow fell in many areas. In Hungary the snow

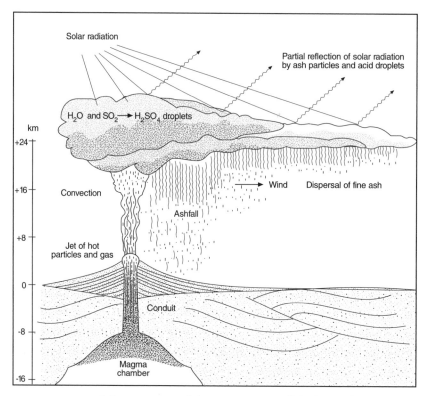

FIGURE 6-3. Schematic view of the 1815 eruption of Tambora. The eruption column is thought to have been at least 25 kilometers high. In addition to covering nearby areas with volcanic ash and pyroclastic flows, the eruption blew vast quantities of water vapor (H_2O) and sulfur dioxide (SO_2) into the atmosphere, where they combined to form droplets of sulfuric acid (H_2SO_4). Veils of these droplets, or aerosols, spread around the world, reflecting enough solar radiation to cause abnormally cold weather.

was colored brown from volcanic dust in the atmosphere. Yellowish and reddish snow fell even as far south as the heel of the Italian boot. Crop failures were widespread. The Alpine regions of France, Switzerland, and Austria suffered the most. Snow and frost came early and stayed late, severely shortening the growing season in 1816 and also, to some extent, in 1817. Attempts to replant summer wheat were frustrated by a lack of seed in state granaries. Tambora's eruption caused what has been called "the last great subsistence crisis in Europe's history."[3]

The bad weather followed closely on the social and economic disruptions of the Napoleonic Wars. Because of wartime casualties, there was a shortage of men to work the farms, especially in France and Germany. Moreover, the fighting had devastated the northern departments of France, the country's breadbasket. The cost of basic foods skyrocketed.

As a result of those converging factors, there was famine in many European countries in 1816, especially in cities. In Paris a record number of people starved to death. Political problems in postwar France had led to a decline in industrial activity, resulting in unemployment in the cities. That and the high cost of food led to local insurrections. Rioting broke out in Poitiers because a tax of three francs was levied on each bushel of wheat, on top of a price that had already doubled. Groups of armed peasants raided farms, looted grain markets, and broke into bakeries. In the Loire valley, grain carts on the way to market, though accompanied by gendarmes, were ambushed and robbed by gangs of peasants. Similar scenes were common throughout France, and there were food riots in other countries as well.

Harvests failed in Ireland, and famine gripped that country. Ensuing health problems led to a typhus epidemic that lasted from 1817 to 1819. The number of deaths is not known but has been estimated at more than 100,000.

—————

In many parts of Europe the sky was overcast much of the time, and rain fell incessantly. In 1816 Lord Byron reflected that melancholy season in a poem he called "Darkness," excerpted here:

> I had a dream, which was not all a dream.
> The bright sun was extinguish'd, and the stars
> Did wander darkling in the eternal space,
> Rayless, and pathless, and the icy earth

Swung blind and blackening in the moonless air;
Morn came and went—and came, and brought no day,
And men forgot their passions in the dread
Of this their desolation. . . .
. . . .
And War, which for a moment was no more,
Did glut himself again;—a meal was bought
With blood, and each sate sullenly apart
Gorging himself in gloom: no love was left;
All earth was but one thought—and that was death,
Immediate and inglorious, and the pang
Of famine fed upon all entrails.[4]

Byron was living in a rented villa on Lake Geneva in 1816. That dreary summer another great English poet, Percy Bysshe Shelley, rented a villa nearby with his wife, Mary. The Shelleys, Byron, and other friends often spent time together. Mary Shelley described the circumstances in the introduction to a book, the first edition of which was published two years later:

> In the summer of 1816, we visited Switzerland, and became the neighbors of Lord Byron. . . . But it proved a wet, ungenial summer, and incessant rain often confined us for days to the house. Some volumes of ghost stories, translated from the German into French, fell into our hands. . . . "We will each write a ghost story," said Lord Byron; and his proposition was acceded to. . . . I busied myself to think of a story,—a story to rival those which had excited us to this task. One which would speak to the mysterious fears of our nature, and awaken thrilling horror. . . . At first I thought but of a few pages—of a short tale; but Shelley urged me to develope the idea at greater length.[5]

What Mary Shelley went on to "develope," of course—inspired by the pervasive gloom from Tambora's dust—was the immortal Gothic horror novel *Frankenstein*.

———

Northeastern North America also experienced disastrous weather in the spring and summer of 1816. Unlike Europe, the region experienced an abnormally dry year. But as in Europe, there were record low temperatures during the growing season, with night frosts and even occasional snowstorms. Cold weather predominated throughout Canada and as far south as New Jersey.

Killing frosts hit parts of New England from June 6 to 11, on July 9, and again in late August. Snow fell abundantly in the northern New England states, and snow squalls reached into Connecticut. In New Bedford, Massachusetts, a town normally protected from weather extremes by its proximity to the ocean, early June brought severe night frosts and strong winds. One Calvin Mansfield of North Branford, Connecticut, wrote in his diary, "Great frost—we must learn to be humble."[6]

Throughout Canada and New England the dryness and cold weather devastated crops, including wheat and the all-important hay and corn, which provided fodder for farm animals. As a result, many animals died during the following winter. While most farmers were largely self-sufficient and ultimately able to cope with the winter of 1816–1817, food shortages developed in the cities and in northern areas, where farming was possible only at subsistence levels.

At St. Johns, Newfoundland, 800 emigrants were sent back to Europe because of food shortages. In northern Vermont and New Hampshire, farmers fed their pigs with fish caught in local streams. Others had mackerel shipped in from New England seaports. Thus 1816 was remembered by some not only as "the year without a summer" but also as "the mackerel year."

Food prices soared in the cities, and many people went hungry. There were, however, no famines like those in Europe. The price of wheat peaked at almost $2.50 a bushel in 1816. (It did not reach that price again until 1972, when much of the American wheat crop went to replace a failed crop in Russia.) Food prices remained high in 1817. Because of the harsh growing season the year before, most farmers had little seed for new crops. In New York City that year, soup kitchens were opened to feed the poor.

The population of the United States had grown rapidly during the seventeenth and eighteenth centuries, and by the second decade of the nineteenth century virtually all the arable land in New England had been settled. Farm families were large, as attested to by the following epitaph, found on a gravestone in Litchfield, Connecticut:

> Here lies the body of Mrs. Mary, wife of Deacon
> John Buel, Esq. She died Nov. 4, 1768, aged 90—
> having had 13 children, 101 Grand-children, 247
> Grate-Grand-Children, and 49 Grate-Grate-Grand-
> Children; total 410. Three Hundred and Thirty Six
> Survived her.[7]

More and more sons were inheriting smaller and smaller farms. To make a living, many took their wives and children and moved to Ohio and the territories farther west. The bad harvests of 1816 and 1817 accelerated that migration, which constituted a significant shift in American population. The stream of new settlers added tens of thousands of inhabitants to the fertile lands west of the Allegheny Mountains.

A poignant reminder of that miserable time in New England, and of the steadfast Yankees who coped with it, is an epitaph on a gravestone in an old cemetery in Ashland, New Hampshire:

> 1771 REUBEN WHITTEN 1847
> SON OF A REVOLUTIONARY SOLDIER.
> A PIONEER OF THIS TOWN. COLD SEASON OF
> 1816 RAISED 40 BUSHILS OF WHEAT ON THIS
> LAND WHITCH KEPT HIS FAMILY AND
> NEIGHBOURS FROM STARVEATION.[8]

No one understood the causes of the cold weather, crop failures, and hunger until much later. Many blamed a decline in morality and a corresponding decrease in church attendance. Some blamed sunspots, others blamed icebergs in the North Atlantic Ocean. Nobody blamed a volcano halfway around the world.

The eruption of Tambora in 1815 devastated the island of Sumbawa; produced famine and disease on nearby islands; disrupted the monsoon and caused famine in India; may have been responsible for a worldwide cholera epidemic; led to cold weather, crop failures, and food riots in Europe; and produced the infamous "year without a summer" in North America. People living then could not know why those disasters happened. They had no way of relating them to a single geological event on a small, remote island in the East Indies.

Sixty-eight years later communications were sufficiently advanced that the worldwide effects of a similar eruption—that of Krakatau, also in the East Indies, in 1883—*could* be related to their cause. It was only then that scientists could begin to answer the questions of 1815.

MOUNT TOBA: BIGGER THAN TAMBORA

The eruption of Tambora in 1815, which may have been the greatest volcanic event in historic times, has become notorious because of its dramatic effect on global weather patterns. But about 74,000 years earlier, another Indonesian volcano, named Toba, exploded in a far greater cataclysm and may have affected human evolution. Toba is estimated to have emitted some 2,800 cubic kilometers of magma, compared with a mere 50 cubic kilometers from Tambora. Its VEI is thought to have been 8, the highest intensity known to date. Volcanologists have adopted the descriptive term *humongous* for Toba's eruption.

That prodigious blast created an immense caldera that now holds Indonesia's largest lake, Danau Toba (Lake Toba). The lake, located in northwestern Sumatra, is surrounded by steep cliffs more than 1,200 meters high. It is 85 kilometers long and has a maximum width of 25 kilometers, and its area is 1,780 square kilometers. The northern end of the lake is as much as 530 meters deep.

Toba's eruption produced an enormous quantity of ash, dust, and volcanic gases. These materials were injected high into the stratosphere, probably reaching altitudes of 30 kilometers or more. Ash from the eruption has been identified in deep-sea cores taken from the bottom of the Indian Ocean and the South China Sea. Evidence from oxygen isotopes in the cores indicates that Toba erupted during the transition from warm to cold climate that initiated the last ice-age glacial cycle. A layer of ash about 30 centimeters thick has been found on the floor of the Indian Ocean some 2,400 kilometers west of the eruption site. Thus Toba's ash cloud must have covered a huge area. Dust and aerosols from Toba would have been carried around the world by high-altitude winds, interfering with incoming solar radiation and leading to global cooling.

Sulfur-containing gases from Toba, upon reaching the stratosphere and combining with water vapor, must have produced vast clouds of sulfuric-acid aerosols (see Figure 6-3). Those aerosols, which spread around the globe and reflected much sunlight, are thought to have lowered average temperatures by at least 10 degrees Celsius. Such a reduction could have led to a "volcanic winter" that may have lasted several years. This abrupt change in global climate came at a time of great stress to early humans, as tribes who had settled in the northern reaches of Eurasia were retreating southward as the ice fields advanced. The harsh climate, which lasted for several years, also affected lower latitudes and sorely taxed those migrating groups and no doubt led to a significant decline in their numbers.

Through DNA studies, scientists have been able to estimate the size of human populations at various times in the past. One finding is that between 70,000 and 80,000 years ago—about the time when Toba erupted—the earth's human population was reduced to possibly only about 10,000, creating an "evolutionary bottleneck." If such a bottleneck did indeed occur, humankind must have come close to extinction. And if the eruption of Toba was related to that bottleneck, it is likely that other "humongous" eruptions in earlier times created similar threats to humankind. The hot springs and geysers in Yellowstone National Park in the United States, for example, are located in an enormous caldera formed by three volcanic eruptions of truly humongous proportions that occurred between about 2 million and half a million years ago and no doubt had global effects. Could those events have been related to the extinction of certain early hominid species in Africa?

7 · Krakatau, 1883: Devastation, Death, and Ecologic Revival

A horrifying spectacle presented itself to our eyes; the coasts of Java, as those of Sumatra, were entirely destroyed. Everywhere the same grey and gloomy colour prevailed. The villages and trees had disappeared; we could not even see any ruins, for the waves had demolished and swallowed up the inhabitants, their homes, and their plantations. . . . This was truly a scene of the Last Judgment.

R. A. Sandick, "Eruption on Krakatau"

IN 1883 KRAKATAU was a small, uninhabited volcanic island in the Sunda Strait between the large Indonesian islands of Java and Sumatra. Long believed extinct, the volcano erupted that year in a series of cataclysmic explosions that were heard for thousands of kilometers in every direction—certainly among the loudest noises ever heard on earth. Enormous quantities of ash and pumice were thrown into the atmosphere, and much of the island collapsed, forming a huge caldron-shaped depression, or caldera. Giant sea waves, or tsunamis, crashed onto nearby shores, destroying more than 160 towns and villages and killing as many as 40,000 people. The eruption was one of the most devastating natural catastrophes in history.*

*The 1815 eruption of Tambora, about 900 miles east of Krakatau on the island of Sumbawa, was more powerful than the eruption of Krakatau, and

After the initial eruption there was an almost continuous fall of volcanic ash for three days. Fine dust and volcanic aerosols were propelled high into the stratosphere and encircled the earth within two weeks. For almost three years, pollution in the air caused a variety of atmospheric effects—spectacular sunsets and sunrises, halos, and bluish-green tints to the sun and moon—in many parts of the world.

Plant and animal life on Krakatau and nearby islands was obliterated by the eruption. Thick layers of gray ash, pumice, and mud covered the ground where once lush jungles grew and wildlife flourished. But a year later grass had begun to grow. Within a few decades the islands were again cloaked in vegetation, and many species of birds, mammals, reptiles, and insects had returned. Thus not only does the eruption of Krakatau in 1883 rank among the world's worst natural disasters, it is considered a classic example of biological regeneration.

The islands of Java and Sumatra are part of the Indonesian volcanic arc, which bends sharply in the region of the Sunda Strait. Java's volcanoes have a predominant east-west alignment, and those on Sumatra trend northwest-southeast (see Figure 7-1). Krakatau, between Java and Sumatra, is one of a series of volcanic islands located above an active fault zone that trends north-northeast—quite different from the alignment of volcanoes on either Java or Sumatra. The most thoroughly studied part of the zone extends from Panaitan Island, off westernmost Java, to the southeastern tip of Sumatra (Figure 7-2, top). Krakatau and its associated volcanoes are much smaller than other volcanoes in Indonesia, but their explosive capability is as great or greater.

The development of the Krakatau fault zone is related to the formation of the Sunda Strait. According to a Javanese legend recounted in *Pustaka Raja* (Book of kings), a volcano called Kapi exploded in 416 C.E. and created a great abyss, which

the death toll was much higher. Global communications were considerably more limited in 1815, however, and Tambora never achieved the notoriety of Krakatau.

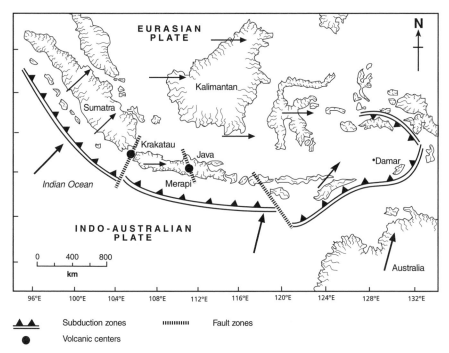

FIGURE 7-1. Tectonic setting of Indonesia, showing the present-day northeasterly drift and subduction of the Indo-Australian plate beneath the eastward-moving Eurasian plate and the resulting volcanic centers of Krakatau and Merapi.

divided a single ancestral island into two parts—today's Java and Sumatra. The legend describes the event as follows:

> A great glaring fire, which reached to the sky, came out of the . . . mountain; the whole world was greatly shaken, and violent thundering . . . took place; . . . the noise was fearful, at last the mountain Kapi with a tremendous roar burst into pieces and sunk into the deepest of the earth. The water of the sea rose and inundated the land. . . . After the water subsided . . . Kapi and the surrounding land became sea and the island . . . divided into two parts.[1]

It is unlikely that a single eruption, as the legend has it, would have created the Sunda Strait. In southeastern Sumatra,

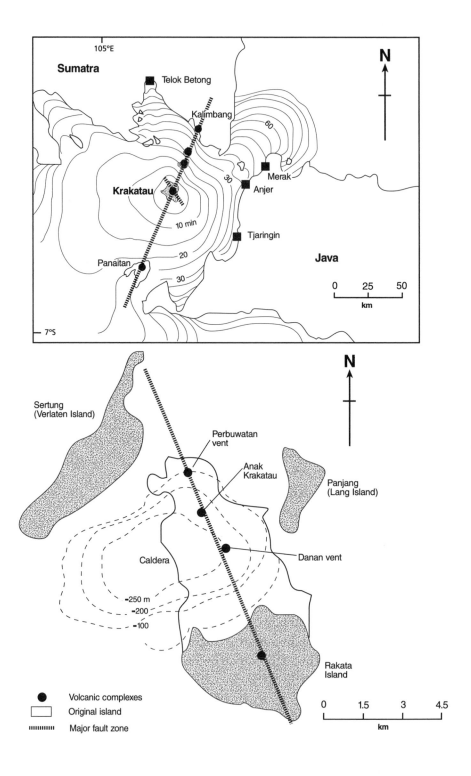

Sumatra

105°E

Telok Betong

Kalimbang

60

Krakatau

Merak

30

Anjer

10 min

Tjaringin

Java

20

Panaitan

30

0   25   50
km

7°S

N

N

Sertung
(Verlaten Island)

Perbuwatan
vent

Anak
Krakatau

Panjang
(Lang Island)

Danan vent

Caldera

—250 m
—200
—100

Rakata
Island

● Volcanic complexes

☐ Original island

⫴⫴⫴⫴ Major fault zone

0   1.5   3   4.5
km

at the northern end of the Krakatau fault zone, there is a plateau of lava that is a million years old. And Panaitan Island and the southwestern part of Java are covered by compacted volcanic ash and pyroclastic debris, some of which is about a million years old and some 700,000 years old. Thus there must have been at least two periods of major volcanism—but both occurred many hundreds of thousands of years ago, not in 416 C.E., as in the legend. Near today's remnant of the old island of Krakatau are two smaller, sickle-shaped islands named Sertung and Panjang. They are parts of the rim of a large submerged caldera, possibly dating from the legendary eruption of Kapi in 416.

Drilling for oil in the Sunda Strait has revealed thick accumulations of sediments, indicating that the region has undergone rapid subsidence in the past few million years. Subsidence of this type typically is associated with extension, or stretching, of the earth's crust, resulting in fracturing. Extension in the strait is caused by differences in the tectonic motions of Java and Sumatra. Java is moving eastward at about 4 centimeters a year, while Sumatra is moving northeast at about the same rate. It is the northward component of Sumatra's motion that is driving that island away from Java and has been responsible for the opening of the Sunda Strait.

These movements result from interaction between the eastward moving Eurasian tectonic plate and the Indo-Australian plate, which is sliding toward the north-northeast beneath Java while thrusting obliquely beneath Sumatra at a rate of 7.5 centimeters a year. As a result, Sumatra is being pushed

FIGURE 7-2. *Opposite: Top:* The Krakatau fault zone and the chain of volcanic islands between Java and Sumatra. Also shown is the reconstructed pattern of tsunami waves that originated during the eruption of 1883 and their rate of propagation. Adapted from Yokoyama, "Krakatau Tsunami." *Bottom:* The Krakatau archipelago with the outline of the original island and submarine caldera that formed in 1883 *(dashed line)*. Note the arrangement of the volcanic centers, which differs from that of the Krakatau lineament shown in the top figure.

sideways and is rotating clockwise. So far that island has rotated 40 degrees with respect to Java. Paleomagnetic evidence shows that about half of the rotation probably has occurred during the past 2 million years. The cumulative amount of rotation has resulted in crustal extension and fracturing between the islands, providing pathways for the ascent of molten rock, or magma, in volcanoes along the Krakatau fault zone.

Before the eruption of 1883 the island of Krakatau consisted of three coalescing volcanoes aligned in a north-northwest direction, like the volcanoes on Sumatra. From north to south the mountains were named Perbuwatan, Danan, and Rakata. Since the alignment of the Krakatau fault zone is north-northeast, however, not north-northwest, Krakatau's volcanism must have developed at the intersection of two major fault zones (compare top and bottom parts of Figure 7-2).

During the 1883 eruption, Perbuwatan, Danan, and half of Rakata collapsed into the emptying magma chamber, creating another submarine caldera.* Enormous volumes of magma and volcanic dust were blown into the atmosphere, and pyroclastic flows filled depressions in the sea floor around the island. Huge tsunamis washed over the shores of western Java and southern Sumatra. Almost half a century later, in 1927, a new volcano, Anak Krakatau (the child of Krakatau), erupted from the submerged caldera. As Sumatra continues to rotate clockwise, the resulting crustal extension will no doubt reactivate the Krakatau fault zone, and the "child of Krakatau" may well repeat the cataclysmic eruptions of its "forefathers."

In 1883, when Krakatau erupted, the islands that today constitute the Republic of Indonesia were controlled by the Nether-

---

*The name *Krakatau* probably is derived from *Rakata*, but it is often spelled *Krakatoa* in English-language writings because someone methodically crossed out *au* and wrote in *oa* wherever the name appeared in a report on the 1883 eruption that was sent to the Royal Society in London. *Krakatau*, however, was the accepted spelling in Indonesia in 1883, and it remains the accepted form there today. In some old Dutch writings the name is also spelled *Craketouw*.

lands and were known as the Dutch East Indies. Indonesia, which extends more than 5,000 kilometers across the equatorial seas between Asia and Australia, comprises Sumatra, Java, Sulawesi (formerly Celebes), most of Kalimantan (formerly Borneo), and many other islands, large and small.

Indonesia has more volcanoes than any other country. More than 130 of them have been active during the past 10,000 years, and 76 have erupted at least once since 1600 C.E. More than a dozen eruptions, of varying degrees of intensity, occur every year. Because of the nation's high population density, those eruptions have caused fully a third of the world's known human fatalities from volcanic activity. The awesome natural forces that have brought death to Indonesians, however, have also brought them life in the form of fertile soils, which, given the warm climate, often allow two or three harvests a year. The soils are continually revitalized by minerals in the volcanic ash from repeated eruptions. As a result, parts of Indonesia, especially Java, are among the most fertile—and densely populated—places on earth.

There were many small villages, or *kampongs,* along the shores of the Sunda Strait in 1883, as well as three sizable towns—Anjer and Merak on the coast of Java, and, in Sumatra, Telok Betong at the head of Lampong Bay. The villagers typically lived in houses framed in bamboo, with walls of woven rattan and roofs of palm thatch. Fishing provided an important source of food. The shores were fringed with coconut palms, and a variety of fruit trees grew farther inland. Cultivated fields occupied most of the coastal lowlands.

In the nineteenth century, ships approaching the East Indies from the west ordinarily sailed through the Sunda Strait between Sumatra and Java. Many stopped at Anjer to obtain supplies and pick up pilots who were familiar with the Java Sea. The small island of Krakatau, in the approaches to the strait, was among the first landmarks sighted after the 8,000-kilometer voyage across the Indian Ocean. Though uninhabited except for a brief time as a penal colony, the picturesque island was a welcome sight, with lush tropical vegetation extending from the seashore to the tops of its three mountains, Rakata, Danan, and Perbuwatan. Easily the most prominent was the cone-shaped Rakata, over 800 meters high. Danan, with an elevation of only about 450 meters, had several peaks, probably representing the remnants of an ancient crater rim. Perbuwatan was little more than a range of hills.

With an area of almost 34 square kilometers, Krakatau was the largest of a group of four islands. Verlaten (deserted) Island lay to the northwest and Lang (long) Island to the northeast. Today Verlaten is called Sertung, and Lang is called Panjang. Between Lang and the northern part of Krakatau was a tiny islet called Poolsche Hoed (Polish hat).

Although Krakatau was known to be a volcanic island, there are no clear accounts of volcanic activity in that area until 1883, except for reports of eruptions in 1680 and 1681. The burning island in Figure 7-3 ("Het Brandende Eiland") may well depict one of those late-seventeenth-century eruptions. By the nineteenth century, the volcano was considered extinct. The first evidence of its awakening was a series of earthquakes between 1877 and 1880. One particularly severe jolt in September 1880 destroyed the upper part of the lighthouse on Java's First Point, the western tip of the island. During the first week of May 1883, a second series of tremors was felt in western Java. Earthquakes are common in the Indonesian archipelago, however, and at that time they were not necessarily considered harbingers of volcanism.

But on May 20, 1883, at about 10:30 in the morning, the crew of the German warship *Elizabeth,* traversing the Sunda Strait, had a close-up view of the first eruption of Krakatau in at least two centuries. The ship's captain reported a cloud of ash and dust that rose almost vertically from Perbuwatan to a height of 11 kilometers. Volcanic ash fell in Anjer, about 60 kilometers from Krakatau. The explosions were heard as far away as Batavia (now Jakarta), 160 kilometers east of the island. By 2:00 P.M. the Sunda Strait region was in darkness. On the twenty-second a fiery glow above Krakatau could be seen from the coast of Java. Earthquake tremors and relatively

FIGURE 7-3. An early etching titled *"Het Brandende Eiland"* ("The Burning Island") by Jan V. Schley, presumably showing an eruption of Krakatau in about 1680. Private collection.

minor detonations continued all through May and June. There were repeated ashfalls, and ships in the Sunda Strait reported sheets of pumice floating on the water in large, coherent masses.

The earthquakes continued intermittently, but the first indication that a catastrophe was in the making came early in the afternoon of Sunday, August 26. It was then that Krakatau exploded with a roar that sent a black, churning cloud of volcanic debris almost 25 kilometers into the atmosphere above the Sunda Strait. Increasingly severe explosions followed one after another, spewing great quantities of ash and pumice into the air. Pyroclastic flows plunged into the sea, triggering a series of tsunamis, the first of which battered the nearby coasts of Sumatra and Java that evening. At the head of Sumatra's Lampong Bay, the town of Telok Betong was partially flooded, and several houses were swept away. The water level rapidly rose and fell a meter or more there and also at Anjer, in Java on the eastern side of the strait, where boats broke loose from their moorings. Houses were destroyed in Tjaringin, a town 35 kilometers south of Anjer. And at Merak, 16 kilometers north of Anjer, the sea washed away a camp for Chinese workers.

Strong earthquakes were felt in the area, and deafening explosions continued all night, keeping people awake as far away as Batavia. Aboard the *Charles Bal,* a British ship sailing past Krakatau that night, the crew frantically shoveled volcanic ash from the decks, lest its weight capsize the vessel. Eruptions continued, and early Monday morning, the twenty-seventh, a tsunami destroyed virtually the entire town of Anjer and many nearby kampongs. One of the very few survivors, an elderly Dutch pilot, gave the following account of his ordeal that morning:

> Looking out to sea I noticed a dark object through the gloom, traveling towards the shore.
> . . . A second glance . . . convinced me that it was a lofty ridge of water many feet high, and worse still, that it would soon break upon the coast near the town. . . . I turned and ran for my life. . . . In a few minutes I heard the water with

a loud roar break upon the shore. Everything was engulfed. Another glance around showed the houses being swept away and the trees thrown down on every side. . . . a few yards more brought me to some rising ground, and here the torrent of water overtook me. . . . I was soon taken off my feet and borne inland. . . . The waters swept past, and I found myself clinging to a cocoanut palm-tree. Most of the trees near the town were uprooted and thrown down for miles, but this one fortunately had escaped and myself with it.

The huge wave rolled on, gradually decreasing in height and strength until the mountain slopes at the back of Anjer were reached, and then . . . the waters gradually receded and flowed back into the sea. The sight of those receding waters haunts me still. As I clung to the palm-tree . . . there floated past the dead bodies of many a friend and neighbour. Only a mere handful of the population escaped. Houses and trees were completely destroyed, and scarcely a trace remains of where the once busy, thriving town originally stood.[2]

Across the strait, on the coast of Sumatra, another great wave struck Telok Betong and stranded two ships, the *Berouw* and the *Marie,* on the beach. Other tsunamis later in the morning completed the destruction of Anjer, Merak, and Telok Betong. Only debris from wrecked houses and boats, and a few trees stripped of foliage, could be seen upon a desolate surface of gray mud where thriving towns had existed before. The wave that struck Merak was powerful enough to rip 100-ton coral-limestone blocks from the sea floor and deposit them on shore. The waves also carried a railroad locomotive almost 50 meters from the track it had rested upon.

All that was mere prelude to the culminating eruption of Krakatau, a stupendous paroxysm of exploding volcanic power on August 27. A series of at least four blasts, starting at 5:30 A.M. and reaching a climax at 10:15 A.M., literally blew up the island. A glowing cloud of smoke, fire, and incandescent ash and pumice roared to a height generally thought to have been about 40 kilometers. The noise was heard as far away as central Australia, the Philippines, Ceylon (now Sri Lanka), and

Rodriguez Island, about 4,700 kilometers across the Indian Ocean (Figure 7-4, top). Most people who heard the sounds in those remote areas mistook them for cannon fire. It has been estimated that the total volume of ash and pyroclastic flows ejected by Krakatau was about 30 cubic kilometers, giving the eruption a VEI (volcanic explosivity index) of 6, considered "huge" by volcanologists and among the highest known in historical times.

Two-thirds of the island collapsed beneath the sea, and thick pyroclastic flows covered the adjacent sea floor and nearby islands. Almost 23 square kilometers of Krakatau, including all of Danan and Perbuwatan, subsided into a caldera about 6 kilometers in diameter (Figure 7-2, bottom). Rakata was literally cut in two, its northern half slipping into the depths, leaving a steep wall that, still today, provides a fascinating cross section of stratification within a volcano. Thick layers of volcanic debris somewhat enlarged the south and west sides of the island.

Immediately after the eruption, at the place where Danan once rose to an elevation of 450 meters, there was seawater to a depth of more than 250 meters. The original depth of the caldera may have been as much as a thousand meters before it accumulated debris from collapsing walls and later volcanic activity. A telegram sent from Batavia to Singapore the next day was both eloquent and poetic: "Where once Mount Krakatau stood, the sea now plays."[3]

Of the other islands in the Krakatau group, tiny Poolsche Hoed was no more. Like most of Krakatau itself, it had collapsed into the caldera. Verlaten and Lang islands, however, remained intact. Verlaten, in fact, had grown to three times its original size, and Lang Island was slightly larger than it had been before. The islands were enlarged mainly by pyroclastic flows that deposited vast quantities of volcanic material in the waters between Krakatau and Sebesi, about 16 kilometers to the north. Where a deep ship channel had been, in fact, there were now shoals and even two new islands. Named Steers and

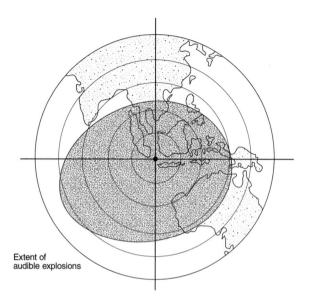

Extent of
audible explosions

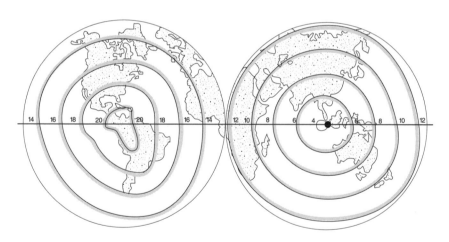

Shock waves circling the earth
(in hours from inception)

FIGURE 7-4. *Top:* The extent of audible explosions during Krakatau's 1883 eruption. *Bottom:* Propagation of the earliest shock waves that circled the globe (in hours from inception). Adapted from Strachey, "Krakatau Airwave," 182.

Calmeijer, the new islands disappeared within a few months as waves eroded the loose material.

Following the cataclysmic blasts of August 27, an immense tsunami smashed ashore on both sides of the Sunda Strait, rampaging inland as far as 4 kilometers in low-lying coastal areas of Java and Sumatra before surging back out to sea (Figure 7-2, top). The waves averaged probably about 15 meters high, but as they pushed into V-shaped inlets and narrow river valleys, their height increased to as much as 40 meters. Those juggernauts swept away whatever was left of Anjer, Tjaringin, Merak, and Telok Betong. In addition they destroyed 165 kampongs and gravely damaged 132 more. The origin of the waves is still in dispute. Many geologists have assumed that the sudden collapse of the volcano and the formation of the caldera were responsible, but the waves also could have been caused by the enormous volumes of pyroclastic debris that slammed into the sea. Most likely both phenomena played a role.

At Telok Betong, the gunboat *Berouw*, which had been beached earlier in the morning, was carried more than 2 kilometers inland and marooned in the valley of the Kuripan River almost 10 meters above sea level. All twenty-eight members of her crew were killed.* The *Marie*, which also had been beached that morning, was found afloat the following day, presumably dragged from the beach by the retreating tsunami. Even in Batavia, the tsunami flooded canals and streets.

Islands within and near the Sunda Strait did not escape destruction. Some were completely submerged by the giant sea wave of August 27. On Sebesi and an uninhabited nearby island, Sebuku, the great wave stripped away all vegetation, leaving only stumps of trees. On Sebesi it destroyed all signs of human habitation. Some 3,000 people were washed into the sea; there were no survivors. In the narrowest part of the

---

*In 1939 one of the authors (Zeilinga de Boer), as a young boy, visited the remains of the *Berouw* with his father. The rusted hulk was overgrown with vines and had become home to a colony of monkeys.

strait, on the Zutphen Islands and an island called Dwars-in-den-Weg (athwart-the-way), the wave destroyed all vegetation up to a height of approximately 20 to 40 meters. Even 80 kilometers east of the strait, the low-lying Thousand Islands were under at least 2 meters of water. The inhabitants saved themselves by climbing trees.

In a resolution of October 4, 1883, the Dutch government appointed a mining engineer and geologist named R.D.M. Verbeek to investigate the nature and consequences of the eruption of Krakatau. His comprehensive report is a classic of geological literature, and it made him famous. It is a point of departure for most subsequent work on the volcano. In his report Verbeek succinctly expressed his impression of that unparalleled catastrophe: "The volcano chose to announce, in a loud voice, to the inhabitants of the archipelago, that although almost insignificant among the many colossal volcanic mountains of the Indies, it yielded to none of them with regard to its power."[4]

No one knows how many people were killed by Krakatau: accurate population figures were not available in 1883, and thousands of bodies were washed out to sea or otherwise never recovered. Nevertheless, Dutch authorities calculated that 36,417 people perished, 90 percent of them killed by the tsunamis. The other 10 percent, who lived downwind from Krakatau along the nearby coast of Sumatra, were burned by hot clouds of volcanic ash that sped across the water on cushions of steam. Ash and dust in the atmosphere cast the entire region into darkness, adding to the terror of a populace battered by thunderous explosions, burning ash, and gigantic waves they could not see bearing down upon them.

The following account, adapted from a report published in the Batavia *Courant* about two weeks after the disaster, describes what one observer found where the town of Tjaringin had been:

> Thousands of corpses of human beings and also carcasses of
> animals still await burial, and make their presence apparent

by an indescribable stench. They lie in knots and entangled
masses impossible to unravel, and often jammed along with
cocoanut stems among all that had served these thousands
as dwellings, furniture, farming implements, and adornments
for houses and compounds.[5]

Such appalling scenes were common along both shores of the
Sunda Strait.

Though the areas inundated by the tsunamis were limited
to coastal lowlands, it was those very areas that supported
most of the population, who depended on agriculture and
fishing for a livelihood. The few survivors were without food
or shelter, their towns and villages gone. Roads and landmarks
had disappeared, washed away or buried in mud. It was impossible to tell where buildings had stood or where property lines
had been. Survivors were in shock, and chaos prevailed. Disease and famine subsequently added thousands of victims to
the volcano's toll.

Many poor people who lived in the hills, and thus had not
been directly affected by the tsunamis, descended from their
villages to look for valuables and to rob any dead bodies they
could find. Robber hierarchies developed, and territorial fighting broke out. The Dutch colonial government sent army and
police units to regain control, but because roads, railroads,
and harbor facilities had been destroyed, the effectiveness of
the units was limited. The anarchy was short-lived, however,
as the lowlands were picked clean after a few months.

Destruction of irrigated rice fields and the removal of soil
by tsunamis rendered large areas infertile for decades. Most of
the few people who survived the catastrophe migrated to
interior parts of the islands. The reefs that provided fishing
grounds had been destroyed, and agriculture was virtually
impossible. Even with intense human effort and the passage
of time, the region never regained the level of prosperity it had
enjoyed before 1883. In fact, the entire Ujung Kulon peninsula
of westernmost Java was later designated a national park.
On an island with one of the highest population densities on

earth, such a designation remains a poignant reminder of the area's continuing unsuitability for human habitation.

For weeks after the eruption, ships in the Sunda Strait, the Java Sea, and the Indian Ocean encountered huge fields of floating pumice, debris, and driftwood, as well as more grisly flotsam. On August 27 the captain of the British ship *Bay of Naples*, in the Indian Ocean some 350 kilometers south of the strait, reported seeing large tree trunks, carcasses of animals, and many human corpses floating by. That same day, the Dutch mail steamer *Gouverneur-Generaal Loudon*, trying to sail from Telok Betong to Anjer, found the passage between Sumatra and Sebuku Island blocked by so much debris that it looked like solid ground. And on December 6, in the middle of the Indian Ocean, the British steamer *Bothwell Castle* encountered floating pumice so thick that it supported several seamen who walked around on it.

Krakatau disgorged so much pumice, in fact, that it choked Lampong and Semangka bays on the Sumatra side of the Sunda Strait and the smaller bays on the Java side. Relief ships were unable to reach Telok Betong for weeks. Floating pumice, worked loose from the bays by storms or high tides, drifted into the Java Sea and the Indian Ocean for months after the eruption. Some of it washed up on the east coast of Africa as much as a year later.

The sea waves that were generated when the volcano erupted radiated out from the Sunda Strait like huge ripples, passing eastward through the Java Sea and westward across the Indian Ocean. Typically, wherever the waves came ashore they were preceded by a withdrawal of the sea. Early in the afternoon of the twenty-seventh near Bombay, India, a sudden drop in sea level stranded fish on the beach, where they were quickly picked up by grateful bystanders. At Auckland, New Zealand, on the twenty-ninth, a 2-meter wave washed a number of vessels ashore. Moving at speeds approaching 500 kilometers an hour in deep water, the waves rounded the tip of Africa and sped up the Atlantic Ocean. They were detected by a tide gauge at Le Havre, France, late in the evening of the twenty-eighth, some thirty-three hours after the climactic eruption of Krakatau.

Krakatau also generated atmospheric pressure waves, or shock waves, that caused considerable damage in nearby areas. Even as far away as Batavia and Buitenzorg (now Bandung), Java, they broke windows and cracked walls of buildings. Barometers showed that the shock waves traveled around the earth as many as seven times, converging at the volcano's antipode near Bogotá, Colombia, and bouncing back as shown in Figure 7-4 (bottom). The first wave reached Bogotá early in the morning on August 28, exactly nineteen hours after its origin on the twenty-seventh. The last passage of a pressure wave from Krakatau was recorded at Washington, D.C., on September 12.

Dust from Krakatau fell as far as 2,500 kilometers downwind in the days immediately following the eruption. The finer particles, propelled high into the stratosphere, remained suspended for years, upper-atmosphere winds near the equator carrying them around the earth in only two weeks. After making a second circuit, the dust cloud spread far to the north and south before finally dissipating.

Also emitted were large volumes of sulfur dioxide, a gas that combined with hydrogen in the stratosphere to form droplets of sulfuric acid. Extensive veils of those acidic aerosols reflected enough sunlight, and its warmth, to cause a global lowering of average temperatures by several degrees. Although less than the worldwide temperature decrease following the eruption of Tambora in 1815, the decrease in 1883 was significant.

Sunlight filtering through the dust particles and aerosols created spectacular optical effects over 70 percent of the earth's surface. For at least three years there were strange colors in the sky, halos around the sun and moon, and extraordinary sunrises and sunsets. The proverbial "blue moon" really existed then from time to time, as well as a blue sun, especially upon rising and setting. And sometimes the sun and moon were observed to be bright green. Magnificent sunsets were reported in many areas, often so brilliant that they suffused the sky with a red glow that was mistaken, in some cases, for a distant fire. Indeed, fire engines were called out in London, New York City, and other cities, the firefighters searching in vain for the conflagrations.

Artists were fascinated by those celestial displays. From 1883 through 1886 William Ashcroft, an English painter, spent many evenings on the banks of the River Thames in London, sketching the ever-changing colors. In 1888 more than 530 of his pastel sketches were exhibited in the South Kensington Museum, now the Science Museum. An American landscape artist, Frederic Church, painted an exceptionally beautiful sunset over Lake Ontario in 1883.*

In 1892 Alfred, Lord Tennyson, published his poem "St. Telemachus," which began with the following lines, suggested by Tennyson's memory of the eruption of Krakatau:

> Had the fierce ashes of some fiery peak
> Been hurled so high they ranged about the globe?
> For day by day, through many a blood-red eve,
> . . . .
> The wrathful sunset glared. . . . [6]

————

The eruption of Krakatau in 1883, though an appalling catastrophe in human terms, provides a fascinating study in biological regeneration. Most authorities agree that the islands of the Krakatau group, as well as Sebesi and Sebuku and some smaller islands nearby, were stripped of all life by the eruption.

*In the twentieth century the motion picture industry became interested in the devastating eruption. Hollywood produced a melodramatic film in 1969 based on the event. Called *Krakatoa, East of Java*, it is about as factual as its title—Krakatau, of course, being *west* of Java. The story involves a ship captain who, while taking on cargo in Singapore, is forced to accept a number of convicts as passengers. The convicts create predictable problems as the captain sails for Krakatau to find a sunken treasure ship while helping his girlfriend find her lost son. They arrive just in time to cope with the eruption and a giant "tidal wave." In 1985 *The Motion Picture Guide* described the film as "almost as much of a disaster as the actual disaster it depicted" (p. 1561).

Krakatau, Verlaten, and Lang islands were devastated by the eruption itself and by the vast quantities of ash and pumice deposited on them by pyroclastic flows, while Sebesi and Sebuku were denuded by the tsunamis that overwhelmed them.

The gradual return of life to the islands amounted to an experiment in natural colonization as plants and animals arrived by air and water. There is some evidence that a few deep-rooted plants may have survived to sprout again, especially on Sebesi, where buried rhizomes were sending up shoots soon after the eruption. But for the most part the islands were a sort of biological tabula rasa, a blank slate upon which a history of new life was to be written. Of primary interest to biologists were the ways in which new flora and fauna arrived on such islands and the sequences in which new plant and animal communities established themselves.

The first to arrive—undoubtedly birds, as well as seeds in bird droppings, spores wafted from the mainland on the wind, or small animals that floated across the water on pieces of pumice or driftwood—found an inhospitable environment of barren mud or deep deposits of still-hot ash and pumice. But eventually plants took root, and larger animals arrived—some by swimming—and survived. This process of renewal is aptly described by the French poet Max Gérard in an untitled verse in *Volcano,* a book published in 1975 by volcanologists Maurice and Katia Krafft:

> And then,
> the most humble of plants,
> a moss.
> And then,
> one morning,
> the first sound of an insect,
> so dry
> you would think it
> still mineral.
> And then,
> hope.[7]

After the eruption, R.D.M. Verbeek, the geologist appointed by the Dutch government to investigate the disaster, was among the first human visitors to what little was left of Krakatau—now a small island comprising only the remaining half of the volcano Rakata. In October 1883 he found no life at all on Rakata. In May 1884 a French mission, sent to study the volcano, also found no trace of plant or animal life except for a single tiny spider, optimistically spinning a web—the first known post-eruption organism living on the island, no doubt carried there from Java, about 40 kilometers distant, by the prevailing winds. But in the fall of 1884 Verbeek reported seeing a few blades of grass that had emerged from the volcanic ash.

The first botanist to visit Rakata after the eruption was Melchior Treub, director of the Botanical Gardens at Buitenzorg. In June 1886, almost three years after the cataclysm, he found 26 species of plants, including mosses, ferns, and grasses—all having seeds that can be spread by winds or ocean currents. The plants grew mostly in isolated patches separated by bare ground. Another investigator found 64 species in 1897, some growing in recognizable plant communities. On the upper slopes of the volcano, there were large grass-covered areas and even a few shrubs and trees. And 108 species were found in 1906, when ground orchids were seen to be flourishing on the steep walls of ravines that had been cut by intermittent streams, and grasses were abundant near the shore. Above the shore grasses there was a narrow zone of mixed forest, and ferns like those Treub had seen at low elevations were now growing higher up on the mountain.

By 1928 the upper slopes of the mountain supported several species of tropical trees and shrubs. Ferns flourished on the shaded walls of ravines. In 1934 W.M.D. van Leeuwen, having succeeded Treub as director of the Buitenzorg Botanical Gardens, reported finding 271 species of plants; he also reported significant changes in the flora (Figure 7-5). On earlier visits, he had found more plant communities, but in 1934, he found that some plants that had flourished earlier had disappeared.

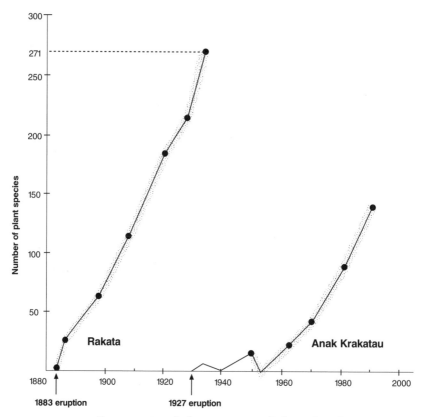

FIGURE **7-5**. Regeneration of plant species on Rakata after the 1883 eruption of Krakatau and on the island of Anak Krakatau, which emerged from the sea in 1927.

Though steppe grasses still covered large areas, there were now many clumps of trees.

Thus there was a gradual increase in the number of plant species, as would be expected with gradually increasing environmental suitability for different species to thrive. Plants of the forest understory, for example, were not able to survive on the island until the first trees had formed clusters dense enough to shield them from the sun. And plants with edible seeds, which are distributed in the excrement of animals, were among the last to be found, for animals could not live on the island in significant numbers or diversity until plant life had developed sufficiently to provide appropriate habitats.

Fig trees played an essential role in the regeneration of plant and animal life—a role that illustrates the complexity of the process. Figs are key components of all healthy lowland tropical forests. It has been found, for example, that in Malaysia, which includes the southern Malay Peninsula and the northwestern part of Borneo (Kalimantan), twenty-nine species of figs provide food for at least sixty species of birds, about a quarter of the forest bird population, and seventeen species of mammals. Figs are so important because they are abundant and they have nontoxic seeds, a wide range of sizes, and a high proportion of edible flesh. Moreover, the ripe fruits are available almost year-round.

After a tropical island such as Krakatau has been devastated by volcanism, the growth of fig trees accelerates forest diversification by providing food for fruit-eating animals, which, in turn, disperse the seeds in their droppings. However, most species of figs can be pollinated only by the females of a particular species of wasp. Unless those wasps are present, the trees will not bear fruit. On Rakata, not only did the fig seeds have to be carried by birds across at least 40 kilometers of open water but pollinating fig wasps had to be wafted to the island as well, at the proper time for pollination. In 1996 Ian Thornton of La Trobe University in Melbourne, Australia, reported on studies that showed the increase in the number of fig species since Krakatau's eruption in 1883. By 1897 four species had established themselves on Rakata. By 1908 three more species had arrived, and by 1920 there were more than fifteen species. The number of fruit-eating birds and bats increased along with the number of figs.

Animal life on Rakata could not be restored completely, of course, until after the restoration of flora on which plant-eating animals feed. But by 1929, according to K. W. Dammerman, director of the Zoological Museum in Buitenzorg, Rakata was home to several species of mammals, birds, and reptiles such as crocodiles, lizards, and snakes, as well as many kinds of insects and invertebrates such as mollusks and worms. Dammerman estimated that the island had recovered up to 60 percent of its original fauna by that time. So little is known of Krakatau's fauna before 1883, however, that an accurate estimate is not possible. There are even house rats on the island today, thought to have been carried to the island on boats when four European families and a number of native workers lived there between 1915 and 1917.

An interesting speculation is whether the revival of Rakata's ecology might give rise to the evolution of new species or sub-species. It is not impossible that two closely related species, reaching the island more or less concurrently, could create a new form by interbreeding. Or aberrant individuals, reproducing on the island, could pass on features that would permanently distinguish their descendants from the original species. Moreover, a species reintroduced to the island might change in order to adapt to different conditions than it was accustomed to on, say, Java or Sumatra. Since Rakata is now a nature reserve with no human habitations, the house rats living there, for example, may well adapt so completely to field conditions that they will develop characteristics of field rats.

Despite Indonesia's ideal soil and climate, the number of plant species on Rakata remained far lower than the number on islands that had not been affected by Krakatau's eruption. Agricultural, hence economic, rebirth in the coastal areas of Java and Sumatra was equally slow. In July 1910 the Batavia *Courant* reported that agricultural production on the western coast of Java had barely reached a third of that before 1883, and by 1927 there was still a shortfall of at least 50 percent. Moreover, reconstruction on Sumatra's east coast lagged behind that in Java. Thus a single volcanic eruption seriously degraded the economy of a large region for several decades. In western Java, the resulting poverty led many people to migrate toward the central part of the island, adding to the already dense human population there.

The biological effects of Krakatau's eruption in 1883 were not limited to the region of the Sunda Strait. Masses of floating pumice formed rafts that drifted with winds and currents as far as Melanesia to the east and Africa to the west. Some of the rafts, as we have seen, were thick enough and strong enough to support men and trees. One such raft, carried some 7,400 kilometers across the Indian Ocean, reached the coast of Zanzibar in July 1884 and deposited human skulls and bones on a beach, where they were found by children from a nearby mission school. Other pumice rafts carried live plants, seeds, eggs, and marine animals. Not unlike Noah's ark, they formed floating ecosystems, which eventually washed up on distant shores and introduced exotic species to new homes.

While Indonesian flora floated to faraway lands bordering the Indian and Pacific oceans, a plant that was to play a major role in Indonesia's economic development arrived by sea from the United States. The sailing ship *Berbice,* en route from New York to Batavia with a cargo of kerosene, also carried a package containing five seedlings of Brazilian rubber trees intended for the Botanical Gardens at Buitenzorg. The *Berbice* sailed down the Atlantic, around Africa's Cape of Good Hope, and northeastward across the Indian Ocean. As she approached the Sunda Strait early in the afternoon of August 26, her captain, seeing darkness and lightning ahead, assumed they were sailing into a storm and gave the order to shorten sail. By 6:00 that evening, hot ash was falling on the ship, burning holes in clothing and sails. The captain, afraid his cargo of kerosene would be ignited, anchored some distance southwest of the entrance to the strait. He proceeded to Batavia on the twenty-eighth, after the eruption abated, and delivered his cargo. The imported rubber trees, as it turned out, became the progenitors of Indonesia's great rubber plantations and eventually brought much wealth to Java and Sumatra.

───────

Forty-four years after the eruption, on December 29, 1927, some fishermen from Java were startled by smoke and steam belching from the sea between Rakata, Verlaten, and Lang islands. Earlier, in June, they had seen gas bubbles rising to the surface in the same area. Geologists placed the site about half-way between where the vents of Danan and Perbuwatan had been. Activity continued, the submarine volcano rapidly building up material on the sea bed, and on January 26, 1928, the rim of a cinder cone appeared above sea level—but waves and currents soon washed away the loose material.

On March 13 the top of the cone was 25 meters below the surface, but by May 18 it was less than 5 meters deep. It formed a small island again on January 20, 1929, and was named Anak Krakatau, the "child of Krakatau." The next morning it had disappeared, but on the twenty-eighth the crater rim reappeared, and the island grew rapidly. On February 1 it was about 12 meters high, and on the ninth its elevation was about 20 meters. By the time the initial series of eruptions ceased in mid-February, Anak Krakatau was a sickle-shaped island about 40 meters high and 300 meters long.

The island disappeared and reappeared several times after that as waves washed away the unconsolidated material of the crater rim and fresh eruptions built the island anew. Finally, in August 1930, it emerged from the sea for good—that is to say, it has remained in existence and has continued to grow from intermittent volcanic activity from that time until the present. In 1960 a new cinder cone appeared in the crater, which had until that time remained submerged except for parts of its rim. That was the first subaerial, rather than submarine, volcanic vent on Anak Krakatau. Today the island has an area of about 10 square kilometers, and the rim of its central crater rises at least 180 meters above sea level.

Like Rakata, Anak Krakatau has provided ecologists with a natural laboratory in which to observe the processes of biological regeneration—with the advantage that Anak Krakatau is a newborn island, on which those processes have been observed from the very beginning (see Figure 7-5). Colonization began there, of course, much as it did on Rakata after the 1883 eruption, with wind-borne seeds and spores. Today most of the island remains barren, but numerous species of plants and animals are found in one small area. Many of them are still struggling for stability. And eruptions of Anak Krakatau, while continually enlarging the island, have periodically destroyed some of the new growth. The area is sufficiently large now, however, that enough life remains to support further growth and regeneration.

––––––––

The explosion of Krakatau in 1883 was a milestone in the science of geology. It not only provided massive amounts of information on the mechanics of volcanoes but also led to greater understanding of calderas and how they might be formed. Krakatau, in fact, is a textbook example of caldera collapse.

The eruption was also significant for meteorologists. They gathered a great deal of data about the earth's atmosphere as they studied the worldwide distribution of Krakatau's dust and analyzed the various optical phenomena. Their work led to a new understanding of atmospheric circulation patterns, including the existence of strong winds high in the stratosphere.

Oceanographers learned much about the formation and behavior of tsunamis and their terrible effects on low-lying shores. And biologists had a rare opportunity to study how life returns to devastated lands and develops on newly created land.

For the first time, in 1883, there were instruments around the world—seismographs, recording barometers, tide gauges—capable of measuring and recording the worldwide effects of a major volcanic event. And communications were sufficiently advanced that word of the eruption spread quickly, so the effects could be correlated with their place of origin. Krakatau demonstrated that the effects of a major geological event can be global, and that the earth's land, sea, and air are interdependent.

The "vibrating string" of Krakatau is short, representing little more than a century, yet it includes cultural changes that lasted for years, economic aftereffects that lasted for decades, and ecological regeneration that has yet to be completed.

THE GHOSTS OF MERAPI

Living on the flanks of frequently active volcanoes has molded the
Javanese view of their world. Traditionally the mountains have been
considered holy by the Javanese people, and their lives have been intri-
cately interwoven with nature and the cosmos. For example, the cloud-
shrouded summit crater of the volcano Merapi, in central Java, is said
in Javanese mythology to harbor a ghost kingdom with a *kraton* (palace),
rulers, soldiers, servants, and farmers. The ghosts herd cattle, work rice
fields, and live in kampongs. Not all the spirits are Javanese. In one
myth, a European, Baron Kasender, was given a place in Merapi's ghost
kraton as the gardener spirit of Merapi. Merapi's kingdom maintains
strong ties with other ghost kingdoms, notably that of the spirit queen
of the southern (Indian) ocean.

These two spiritual kingdoms are said to be connected by Java's
Progo River. From time to time, spirits called *lampors* are said to drive
horse-drawn carts through the water, creating great turbulence as they
travel upstream or down to visit one another. The *lampors* are personi-
fications of the volcanic mudflows that intermittently course down the
river valley, and also of the tsunami waves that occasionally surge upstream in the river's estuary after offshore earthquakes.

Rather than viewing volcanoes exclusively as feared agents of destruction, as in the myths of Hawaii, the Javanese recognize the beneficial aspects of volcanism. According to the mythology, the ghost king
of Merapi and the queen of the ocean have a loving sexual relationship.
Eruptions of Merapi are seen as spectral ejaculations, the ejaculate
flowing downstream to fertilize the ocean queen. This myth reflects an
awareness that an influx of volcanic ash and mud fertilizes coastal waters
and increases the yield of fish and other seafood.

In Hawaii, by contrast, the natives, in awe of their volcanoes, created the myth of Pele the fire goddess, who, when offended by mortals,
causes the mountains to erupt streams of molten lava, and who stamps
her foot to make the earth shake. The ocean, too, can be destructive
when the sea goddess Namaka o Kahai, Pele's hated sister, sends great

sea waves (tsunamis) ashore to quench Pele's fires or erode the land created by Pele's lava flows.

The contrasting volcanic myths reflect differences in the geology and geography of Hawaii and Java. When a volcano erupts in Hawaii, lava flows commonly reach the ocean, where the molten rock reacts explosively with seawater—the eternal battle between Pele and Namaka o Kahai. On Java the volcanoes are inland, a considerable distance from the sea. Though eruptions frequently are devastating, their lava flows seldom reach the sea, so there are very few encounters between molten lava and seawater. Thus Javanese mythology focuses on the benign, nourishing relationship between mountain ghosts and ocean spirits.

8 · The 1902 Eruption of Mount Pelée: A Geological Catastrophe with Political Overtones

*There is nothing in the activity of Pelée
that warrants a departure from St. Pierre.*

The Governor's Commission of Inquiry, May 5, 1902

ON MAY 8, 1902, Mount Pelée, a volcano at the northern end of the Caribbean island of Martinique, erupted and destroyed the city of St. Pierre, only about 6 kilometers away. The city's entire population—30,000 people—died almost instantly. The volcano had provided many warnings of an impending eruption, if only the science of volcanology had been far enough advanced for the warnings to have been understood. And tragically, for political reasons, government authorities on Martinique discouraged, and in at least one town prohibited, the evacuation of people despite obvious indications of volcanic activity. The catastrophe also ended a political movement that could have led to a more representative form of government on that island colony of France.

The 1902 eruption of Pelée was the first recognized example of what French geologists named a *nuée ardente* (glowing cloud), more widely known today as a pyroclastic flow—a gravity-controlled cloud of superheated gases and fragmental (pyroclastic) material moving rapidly over the surface of the ground rather than upward into the atmosphere. Geologists concluded that the phenomenon was caused by a plug of molten rock, or magma, that had solidified and sealed the volcano's conduit, so the enormous pressures building up within the mountain burst out horizontally for the most part instead of vertically. Such eruptions, not identified before 1902, are now called Peléan eruptions. Interest in the origin of the 1902 event changed the study of volcanoes from a minor branch of geology to an important science in its own right.

After the initial devastating blast, new eruptions occurred throughout the month of May and continued intermittently for three years, until 1905. The volcano was again intermittently active from September 1929 to December 1932 but has been dormant since then. The cumulative thickness of deposits from the 1902–1905 and 1929–1932 events varies from a few meters on the steep slopes near the top of Pelée to several tens of meters in valleys farther down the mountainside.

———

Martinique forms part of the gently eastward-curving volcanic island arc of the Lesser Antilles, an island chain that extends from Grenada northward to the tiny island of Saba, a distance of about 850 kilometers (Figure 8-1). Volcanism in the islands is related to the collision of the Caribbean tectonic plate with a segment of the Atlantic sea floor that is part of the North American plate. The convergence began when new crust started forming along the Mid-Atlantic Ridge, spreading the sea floor and opening the Atlantic Ocean. In the early stages, the Caribbean plate moved northeast, then east, as the floor of the Atlantic spread westward.

At a rate of about 2 centimeters a year, the North American plate is being subducted beneath the Caribbean plate at an angle of 50 to 60 degrees. The subducted slab is now about 140 kilometers beneath Martinique. The magma that finds its way upward into Mount Pelée is believed to have originated in a wedge of the earth's upper mantle that lies between the subducted slab and Martinique's crust.

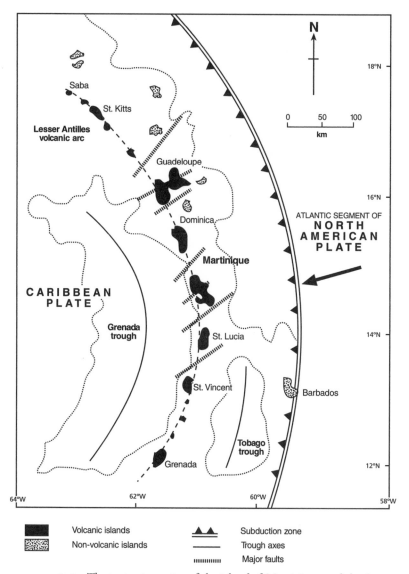

FIGURE 8-1. The tectonic setting of the island of Martinique and the Lesser Antilles volcanic arc, which results from the collision between the Caribbean plate and the oceanic segment of the North American plate.

Partial melting of mantle material in the wedge has formed batches of magma that have ascended and accumulated at the mantle-crust interface at a depth of about 30 kilometers. From there the magma has moved upward into the crust, which contains older volcanic rocks and layers of sedimentary rock. The magma has risen intermittently and is thought to have accumulated first in a large chamber about 15 to 20 kilometers deep in the crust. There it would have become differentiated chemically. Magma rich in heavy elements would have settled to the bottom of the chamber while silica-rich, hence lighter, magma would have accumulated at the top. Increasing pressures in the upper part of the chamber would make the volcano erupt explosively—most likely when fresh batches of magma were being injected into the chamber from below.

During the period from about 16 million to 5 million years ago, volcanic eruptions in the region changed from submarine to subaerial and thus began to form islands. On the island of Martinique, the oldest volcanoes—Morne Jacob, the Pitons du Carbet, and Mount Conil—first erupted 3 to 4 million years ago (Figure 8-2, bottom).

Mount Pelée is believed to have first erupted about 200,000 years ago. Now rising to a height of 1,397 meters and occupying about 120 square kilometers of northern Martinique, it dominates the topography of the island. Erosion from rainfall and runoff, as well as from landslides, has destroyed the once-conical symmetry of the mountain and has left deep valleys in its flanks. Volcanic rocks exposed in the valleys and sampled in drill cores indicate that there were major eruptions around 65 B.C.E., 280 C.E., and 1300 C.E., with many minor eruptions in between. Mount Pelée is among the most active of the Lesser Antilles volcanoes. There were at least forty-six eruptions in the period from 50,000 to 3,000 years ago, and there have been twenty-six since then.

A record of worldwide eruptions compiled by geologists Tom Simkin and Lee Siebert of the Smithsonian Institution in Washington suggests that Pelée generally has erupted at intervals of 50 to 150 years.[1] In historical times Pelée was active in

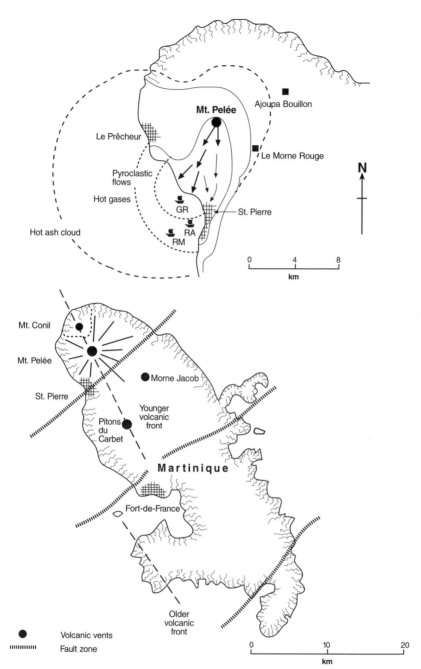

FIGURE 8-2. *Top:* Pelée's eruption of 1902 and approximate extent of the ash cloud, hot gases, and pyroclastic flows. Ship symbols: GR, *Grappler;* RA, *Roraima;* RM, *Roddam. Bottom:* The island of Martinique and the location of Mount Pelée.

1792, 1851, 1902–1905, and 1929–1932. The activity in 1792 and 1851 was minor compared with the eruption of 1902, but it destroyed all vegetation at the top of the mountain—hence the name Pelée, French for *bald* or *peeled*. Large eruptions, like those that occurred from 1902 to 1905, probably occur only once in about 500 years. Even the 1902–1905 events, though disastrous in their consequences for the human population, were relatively minor in terms of geological significance—except for the historical fact that they produced the first eruption identified as a *nuée ardente,* or pyroclastic flow.

The remains of three craters are visible at the higher elevations of Pelée. The largest and oldest was the eruption center up until about 40,000 years ago. Two smaller craters were formed along a northeast-southwest-trending fault inside the older crater. One of those craters, toward the southwest and known in 1902 as L'Étang Sec (the dry pond), had at one time contained water; the other, toward the. northeast, was called the Lac des Palmistes and did contain water in 1902. The topography of all three has been changed considerably by twentieth-century volcanism, and only the two smaller craters are easily distinguishable now.

The first sign of the volcanic activity that destroyed St. Pierre came as early as 1898, when sulfurous gases were emitted from fumaroles, or vents, in L'Étang Sec. Fumaroles also appeared in the upper reaches of the Blanche River, which flowed down the southwest flank of Pelée in a deep gorge marking the trace of the northeast-southwest-trending fault. That activity waned but had resumed by 1901, when there were minor eruptions of steam from superheated groundwater. By February 1902 the gases, carried downwind, were detected near Le Prêcheur, a fishing village on the west coast of Martinique about 6 kilometers from L'Étang Sec (Figure 8-2, top).

The steam and gas emissions, though no one knew it then, indicated that magma was slowly rising into the volcano's conduit, fracturing the surrounding rock. Several low-magnitude earthquakes preceded and accompanied that activity. Their intensity gradually increased, and by the end of April they were occurring frequently.

During the next phase of the eruption, a dome of viscous lava quietly grew within L'Étang Sec as small batches of magma bypassed the volcanic plug and rose to the surface. Meanwhile the great mass of magma that was entering the volcano's conduit from below slowly, relentlessly, forced the plug upward. Roughly cylindrical and some 200 meters in diameter, the solid plug rose through the lava dome and formed what came to be known as the spine of Pelée (Figure 8-3). The spine continued to rise 3 to 6 meters a day, crumbling along its edges but supported by magma extrusions around its periphery. At the time of the catastrophic eruption on May 8, it had reached a height of 115 meters above the former floor of the crater.

Magma forced its way upward in increasingly larger volumes, branching into fissures that radiated from the central conduit, causing inflation, or swelling, of the mountain. On May 4 one such fissure opened beneath the village of Ajoupa-Bouillon on the northeast flank of Pelée, emitting scalding steam and boiling mud and killing a number of people.

Loud detonations continued from the mountain at irregular but short intervals, accompanied by dense smoke and lurid flashes of fire and lightning, reflected in the clouds shrouding the volcano's peak. Especially at night the spectacle was frightening.

Earthquakes, ashfalls, and warming of the earth caused by rising magma drove most ground-dwelling animals down from Pelée's higher elevations. Columns of ants and large numbers of centipedes and snakes (many of them poisonous) invaded outlying plantations and parts of St. Pierre. Birds were overcome by volcanic gases in the atmosphere and died, dropping out of the air like stones.

On May 7 St. Pierre was shaken by detonations like artillery fire that were heard throughout the Antilles. The plug in Pelée's throat had apparently risen high enough for the magma to begin bypassing it explosively.

FIGURE 8-3. The "spine" of Pelée before it crumbled about a year after the eruption. Note the size of the human figures in the foreground. From Lacroix, *La Montagne Pelée et ses eruptions*, plate 1.

At 8:02 A.M. on May 8, there were four deafening explosions in rapid succession. The mountain's climactic eruption began with a vertical column of smoke and ash, which rose from L'Étang Sec and spread across the sky. The eruption was followed almost immediately by a powerful lateral blast from a notch in the southwestern rim of the crater. The blast was, of course, a pyroclastic flow. Comprising dense, brownish-black clouds of superheated gases, incandescent droplets of fresh magma, and fragments of rock from earlier volcanic events, the blast originated either from the rapid collapse of the rising eruption column or from failures in the flank of the volcano—or perhaps from both—and it headed straight for St. Pierre (Figure 8-2, top).

The pyroclastic flow expanded both vertically and horizontally as it sped down the mountainside toward the coast. It set fire to forest trees and cane fields over a wide area. In seconds it engulfed the village of St. Philomène, a northern suburb of St. Pierre, and almost simultaneously it destroyed the city itself. The hot, turbulent cloud is estimated to have been moving as fast as 500 kilometers an hour when it reached the sea.

Pelée erupted again in 1904, and a pyroclastic flow like that of 1902 was photographed by Alfred Lacroix, a French geologist. His photograph, reproduced in Figure 8-4, dramatically shows the height, density, and massive turbulence of the cloud produced by such a catastrophic event.

Magma continued to push upward after the 1902 eruption, and by the end of May the lava dome had risen above the crater's rim. Pelée's spine continued to rise as well, reaching a height of 150 meters above the former crater floor by the end of May. Compared with destructive eruptions of other volcanoes, the eruption of Pelée produced a small volume of magma, probably less than half a cubic kilometer. Nevertheless, because of the height of the eruption cloud, the volcanic explosivity index (VEI) has been estimated at 4, which volcanologists consider "large."

The first pyroclastic flow, on May 8, left a denuded, wind-blown wasteland, crumbled ruins, burned-out shipwrecks, and approximately 30,000 human corpses. There was little left to be destroyed by the subsequent eruptions, which mercifully blanketed the cruel scene with additional layers of ash. The volcano produced many smaller pyroclastic flows from late

FIGURE 8-4. This pyroclastic cloud, photographed during an eruption of Mount Pelée in 1904, presents a graphic idea of what the catastrophic eruption of 1902 must have looked like. From Lacroix, *La Montagne Pelée et ses eruptions*, plate 14.

1902 until July 1905, mostly confined to the original valley of the Blanche River.

Although the eruption of May 8 destroyed the city of St. Pierre and its environs and took a terrible toll in human lives, the devastation was restricted to the northern end of Martinique. Of the island's 970 square kilometers, only about 30 square kilometers were seriously affected.

Martinique was discovered by Christopher Columbus in 1502, but the island was not settled by Europeans until 1635. In that year the French Compagnie des Isles d'Amérique took possession of Martinique for the cultivation of cotton and tobacco. Sugar cane was introduced in 1650, and coffee in 1723. Slavery was instituted early in the eighteenth century because of a growing need for cheap labor to operate the plantations economically. During the course of colonization, the native Caribs were either killed or assimilated, primarily into the black slave population.

The island was occupied by the British for brief periods in the eighteenth and early nineteenth centuries but finally was restored to France in 1814. Slavery was abolished in 1848, whereupon all the island's inhabitants became French citizens. Indeed, a black man named Amédee Knight, who had been educated in Paris, was elected senator in 1899 and for several years represented Martinique in the French national assembly. The island became an overseas *département* of France in 1946.

The capital of Martinique, Fort-de-France, has a population today of about 100,000. In 1902, however, its population was less than 18,000, and St. Pierre, then a thriving seaport with about 26,000 inhabitants, was the most important city on the island. It was from there that the island's exports, mainly sugar and rum, were shipped to Europe and the United States. Often called the "Paris of the West Indies," St. Pierre was the commercial, educational, and cultural center of Martinique. An attractive city, it boasted a renowned school, a cathedral, a theater, and a military hospital, as well as banks, warehouses, factories, and many rum distilleries. St. Pierre had no natural harbor. The city was situated on an open bay, or roadstead, along 3 kilometers of gracefully curving beach. Tiers of sturdy, off-white, red-roofed masonry buildings climbed toward the verdant foothills of Mount Pelée, which provided a picturesque backdrop when St. Pierre was approached from the sea.

At the time of Pelée's eruption in 1902, Martinique had a rigid, well-defined social structure. There was an "upper class" comprising white settlers of French ancestry known as *békés*. Most were plantation owners, who largely controlled the financial and political affairs of the island. A "middle class" consisted of white and mixed-race merchants and their employees, who lived in the towns, and owners of small farms in the countryside. At the low end of the social scale were black and mixed-race plantation workers, laborers, household servants, washerwomen, and *porteuses,* remarkably strong women who worked on the waterfronts of St. Pierre and Fort-de-France.

For centuries the politics of Martinique had been dominated by the conservative *békés,* who advocated white supremacy and had traditionally elected the senator and two deputies who represented the island in government councils in Paris. But in the election of 1899, a new socialist party gave voice to the island's black and mixed-race majority. It was then that Amédee Knight was elected senator, for the first time giving the blacks and those of mixed race their own representative in Paris. Knight had been a successful advocate of reforms in employment, education, and housing and had done much to make the socialists a viable force in the affairs of Martinique.

In 1902 the socialists were on the threshold of wresting political control of the island from the conservative *békés.* Elections had been scheduled for that year—a primary election on April 27 and final balloting on May 11. There can be no doubt that a socialist victory not only would have drastically changed the political spectrum of St. Pierre and therefore of Martinique but also would have challenged the status quo in other French possessions as well. The pre-election tension was exploited by the local newspaper *Les Colonies,* which supported white supremacy. First-page articles on April 21, 22, and 23, headlined "M. Knight et la Guerre des Races," depicted Knight as inciting a racial war.[2]

The socialists were divided between two candidates, but they trusted that no matter which candidate won the primary election, their voters would unite behind him in the final balloting on May 11 and ensure a socialist victory. Sure enough, in the April primary the two socialist candidates together outpolled the conservative candidate. The *békés'* candidate received

4,496 votes,* while between them the two socialists received 4,920 votes.[3]

The pending final election led Martinique's governor, Louis Mouttet, to discourage the evacuation of St. Pierre in spite of increasingly threatening activity of Mount Pelée during the months preceding the disastrous eruption of May 8. A large majority of the conservative voters lived in St. Pierre, and Mouttet surely would not have wanted to see any of them leave the city.

The governor's thinking was perhaps influenced by the fact that the volcano had been slumbering for half a century. Pelée had last erupted in 1851, producing a column of ash that fell on St. Pierre but was soon washed away by a rainstorm. Little was thought of it. Indeed the islanders, over the years, had come to think of Pelée as more of a benign presence than a threat.† They were proud of their mountain. It was a favorite destination for social outings. People would hike to the top to view the placid waters of the Lac des Palmistes. Some of the black population of Martinique had even come to regard Pelée as something of a protector.

Then on April 23, 1902, there was an earthquake that was strong enough to make dishes fall from shelves in St. Pierre. There was a minor eruption and a slight ashfall on the twenty-fourth, and people remarked on the sulfurous odor. The mayor, Roger Fouché, in an effort to forestall any apprehension on the part of the citizens, declared sulfur beneficial to health. Nevertheless, many people were becoming concerned about Pelée's activity. The wife of Thomas Prentis, the United States consul

---

*Curiously, there were only about 4,000 *békés* in St. Pierre. Assuming that half to three-quarters of them must have been women or children, unable to vote, either a very large number of blacks voted for the white conservative (not likely) or some questionable activities attended the balloting.

†It should be pointed out, too, that evacuation orders can backfire. In July 1976 a volcano on the island of Guadeloupe began emitting clouds of steam and ash. A quarter of the island's population, 72,000 people, was evacuated. Intruding magma remained at depth, however, and there was no major eruption. The evacuation had severe economic repercussions. Sugar cane was not harvested, and tourists avoided the island.

in St. Pierre, wrote to her sister in Massachusetts about events during that last week of April:

> The city is covered with ashes and clouds of smoke have been over our heads for the last five days. The smell of sulphur is so strong that horses on the street stop and snort, and some of them are obliged to give up, drop in their harness and die from suffocation. Many of the people are obliged to wear wet handkerchiefs over their faces to protect them from the fumes of sulphur.[4]

On April 30, people in St. Pierre felt three strong earthquakes, probably caused by strain release along the fault beneath the volcano. With the coming of May, the tempo of activity increased. Ashfalls became heavier. Early on May 2, during the night, volcanic ash blanketed the countryside like a snowstorm. Some roads were blocked by fallen tree branches, broken off by the weight of the ash that coated them. In St. Pierre, carriages made no noise that morning as they rolled through the powdery stuff on the city's cobblestone streets. Sulfurous fumes made breathing difficult and led to sore throats and smarting eyes.

Within L'Étang Sec a high cinder cone had appeared. It gushed hot water and had formed a lake in which full-grown trees were submerged. Streams radiating from the summit of Pelée increased in volume, both from rainfall caused by volcanic particles seeding the clouds that normally clung to the mountaintop, and from groundwater forced to the surface by rising temperatures within the mountain.

Mudflows swept down the riverbed on May 3, filling the valley with mud and rocks, some of enormous size. The flows originated as landslides in the upper valley of the Blanche River, probably triggered by seismic activity. Their force and volume were augmented when the water in L'Étang Sec broke through the notch in the crater rim. Mudflows also descended into Le Prêcheur, about 10 kilometers northwest of St. Pierre. The mudflows indicated a general weakening of Pelée's southwestern flank.

Les Colonies had organized a public outing to the top of Mount Pelée, scheduled for May 3, for a picnic and a look at the crater there. Its purpose was to demonstrate that all was well, but the trip had to be canceled because of continuing ashfalls and offensive fumes in the air. Indeed, several people choked to death in St. Pierre that day, and domestic animals were dying on the city streets and in the surrounding countryside from inhaling the sulfurous ash. It was on May 3, too, that the mudflows destroyed Le Prêcheur, killing a number of people. That same day the spinelike protrusion was first seen rising from L'Étang Sec.

Meanwhile a number of rivers on the flanks of Pelée were overflowing their banks. Normally, except during the rainy season, they were small streams that made their way unobtrusively down the mountainside through lush farmland. But now, fed by heavy rains and by groundwater gushing from fissures that had opened in the sides of the mountain, they carried uprooted vegetation, whole trees, and dead farm animals. One such river, the Roxelane, flowed through St. Pierre and flooded low-lying parts of the city before washing its grisly cargo out to sea.

Pelée was sending its warnings, but they went unheeded by the great majority of people in St. Pierre. They had been lulled into complacency, both by imperfect scientific knowledge of how volcanoes can behave and by government officials concerned about an election. A few hundred people did leave the city for Fort-de-France, about 20 kilometers south, either by boat or along the one overland route, La Trace. Their numbers were necessarily limited, however: the available boats were small and could carry but few passengers, and La Trace was just that—a "trace," or primitive track. Moreover, only the wealthy had money for boat passage or carriages to carry them to Fort-de-France. The number who fled was more than offset by refugees from the devastated countryside: the refugees poured into the city, swelling its population to an estimated 30,000. Food was becoming scarce, and there were cases of looting in St. Pierre. Ships approaching the port had to pick their way through masses of debris that had been washed down by the rivers and were floating in the roadstead.

Governor Mouttet finally was stirred to action. On May 4 he sent a cablegram to the minister of colonies in Paris announcing, for the first time, that Mount Pelée had erupted—but that "the eruption appears to be on the wane." And he appointed a commission of inquiry. Of its five members, only one was a scientist—Gaston Landes, a teacher of natural sciences at St. Pierre's *lycée* (high school). Their mission was not to investigate the behavior of Pelée itself but to determine how long the city could "stand the strain of the ash and sulfur" and "whether the terrain around St. Pierre would shield the city from the danger of a lava flow."[5] The governor made it clear that the commission's role was merely to confirm the safety of St. Pierre.

Although May 4 was a Sunday, the *porteuses* on the waterfront of St. Pierre were made to load sugar into a ship waiting to leave for France. The ship was departing unexpectedly early, as her captain was concerned about the behavior of Mount Pelée. The sugar had come from the Blanche River refinery, which was owned by an influential *béké*, Eugène Guérin. But on Monday the fifth, in reprisal for having been forced to work on Sunday, the *porteuses* went on strike. Their refusal to work underscored the burgeoning political power of the socialists on Martinique.

That same day, more mudflows poured down the stream valleys on Pelée's southwestern flank. Some were so wide and voluminous that they swept away plantations, factories, cattle, and people. When they reached the coast and entered the sea, the displaced water created sizable waves that swept down the coast and flooded the St. Pierre waterfront. That evening submarine landslides, probably triggered by the mudflows, ruptured the telegraph cable from Fort-de-France. One end of the cable was later picked up 15 kilometers west of St. Pierre at a depth of 700 meters.

Also on May 5, swarms of small yellow ants known as *fourmis-fous* and large black centipedes called *bêtes-a-mille-pattes* descended the slopes of Mount Pelée like a biblical plague. At the Guérin refinery, the venomous arthropods created havoc as they attacked workers in the cane fields and even invaded the owner's household. St. Pierre itself was invaded by hundreds of snakes, many of them the deadly pit vipers known as fer-de-lances.

Meanwhile back at the Guérin refinery, a greater tragedy was taking place. Among the mudflows of May 5 was one that coursed down the valley of the Blanche River, burying the refinery and the estate under a thick layer of boiling mud. The owner miraculously escaped, but his entire family and many employees were killed. And in St. Pierre, ash-clogged sewers and polluted drinking water contributed to an outbreak of a fatal, highly contagious pox called *la verette*.

By now Pelée had claimed more than 600 lives, and the city was in a state approaching chaos; yet on May 5 the commission of inquiry, fulfilling the governor's wishes, blandly reported that "there is nothing in the activity of Pelée that warrants a departure from St. Pierre."[6]

While all that was happening on Martinique, a similar disaster was in the making 130 kilometers to the south, on the British-owned island of St. Vincent. A volcano called La Soufrière erupted early on May 6 with Pelée-like pyroclastic flows that killed almost 1,600 people and devastated much of the island's farmland. When the inhabitants of St. Pierre awoke that morning, they found their city covered with a thin layer of volcanic ash—ironically not from Mount Pelée but, unknown to them, from a volcanic cloud that had drifted over Martinique from St. Vincent.

Tragically, the catastrophe on St. Vincent compounded the disaster on Martinique. The world knew nothing of what was happening on Martinique because by May 6 seismic activity spawned by the volcano had severed all cable connections with other islands. Thus, in those pre-radio days, Martinique was cut off from the rest of the world.

St. Vincent's communications were intact, however, and authorities on that island sent word of La Soufrière's eruption to England, which promptly dispatched ships to aid the stricken colony. Those ships, and others, passed through ash clouds drifting over the sea from Mount Pelée, but the ash was assumed to have come from La Soufrière. The world remained ignorant of the tragedy of St. Pierre until many hours after the city had been destroyed two days later.

Meanwhile Pelée was roaring continuously, and red-hot rocks (volcanic bombs) were being ejected from L'Étang Sec. Some of them landed in the part of St. Pierre that was nearest the volcano, and houses were set afire. And the volcano was still sending forth great clouds of ash and cinders. The ash coated everything in St. Pierre, and authorities encouraged the citizens to wash down the roofs and walls of buildings. As a result, the ash in the hilly streets turned into a slippery sludge, and people could get about only with great difficulty. Many did not even go into the streets. A strange lethargy had come over the populace, as if they were in a state of shock from all they had endured.

The next day, May 7, *Les Colonies*, still supporting the governor's contention that the volcano was no threat to St. Pierre or its inhabitants, published a story written by the editor but presented as an interview with Gaston Landes, the science teacher. The gist of the story was that St. Pierre was in no danger from Mount Pelée. Governor Mouttet, with his wife, journeyed from Fort-de-France to St. Pierre that day, feeling that their presence would give the people confidence. They died there the next morning, along with thousands of others. Their bodies were never found.

There has been disagreement about Governor Mouttet's role in the tragedy of St. Pierre. Some writers in recent years have argued that Mouttet's visiting the stricken city demonstrates an honest conviction that Mount Pelée presented no serious threat. And others have questioned whether the conservative element in St. Pierre really was so concerned about the election that, despite the threat of volcanic activity, they wanted to keep people in the city until after the election on May 11. But in 1989 the French scholar Solange Contour wrote that the minister of colonies in Paris had ordered Mouttet to keep St. Pierre's voters in town until the election was over.[7]

May 8 was Ascension Day in 1902, and early that morning many of St. Pierre's citizens, most of whom were Roman Catholic, made their way through ash-laden streets to services in the cathedral. Indeed, many had taken refuge there in the preceding days, both out of fear and to escape the ash and cinders raining down upon the city. They were there at 8:02 A.M., when their world abruptly came to an end. At that instant the first pyroclastic flow hurtled down the valley of the Blanche River, then spread out, and, in seconds, engulfed St. Pierre. With the force of a hurricane, the superheated cloud destroyed buildings, twisted iron girders, filled the streets with rubble, and set the city ablaze. Estimates of its temperature vary widely, but most suggest about 900 degrees Celsius when the cloud left the crater and 200 to 400 degrees when it reached the coast. Pieces of melted bottle glass and fused metal in St. Pierre indicate temperatures as high as 1,000 degrees, but such temperatures resulted not from the pyroclastic flow but from extremely hot fires fed by exploding casks of rum.

Not everyone in St. Pierre was killed immediately, though mercifully many were. It became apparent later that many people had died with agonizing slowness from injuries or suffocation or from inhaling the hot gases. According to most reports, the only survivor within the city itself, excluding nearby areas and ships in the bay, was a black man named Auguste Ciparis, who had been jailed after getting into a fight. He had been sentenced to a short term in a dungeonlike cell in the local prison. There, ironically, he was protected from the worst of the hot blast by volcanic ash that had piled up and blocked the cell's one tiny window. He was badly burned, however, and spent four days without food or water, in an agony of ignorance about what had happened to the world outside. Finally his cries were heard, and he was rescued. Ciparis's sentence ultimately was suspended, and he spent the rest of his life touring the United States with the Barnum and Bailey Circus, where he made a living as an exhibit in a sideshow replica of his prison cell.

In the roadstead the hot cloud sped across the water on a cushion of steam, its force overturning and sinking several ships lying at anchor. Ships that did not founder caught fire. Most people aboard the ships met horrible deaths from suffocation or burning. There were some survivors, however. Eventually they were rescued by ships sent from Fort-de-France to find out why telegraph communication between the two cities had suddenly ceased at 8:02 that morning.

The horror of the event was later described by Charles Thompson, assistant purser of the steamship *Roraima:*

> There was a constant muffled roar. It was like the biggest oil refinery in the world burning up on the mountain top. There was a tremendous explosion about 7:45 o'clock. . . . The mountain was blown to pieces. . . . The side of the volcano was ripped out, and there was hurled straight toward us a solid wall of flame. It sounded like thousands of cannon.
>
> The wave of fire was on us and over us like a lightning flash. . . . I saw it strike the cable steamship *Grappler* broadside on and capsize her. From end to end she burst into flames and then sank. The fire rolled . . . down upon St. Pierre. . . . The town vanished before our eyes. . . .
>
> The blast of fire from the volcano lasted only a few minutes. It shriveled and set fire to everything it touched. Thousands of casks of rum were stored in St. Pierre, and these were exploded by the terrific heat. The burning rum ran in streams down every street and out to the sea. This blazing rum set fire to the *Roraima* several times. Before the volcano burst the landings of St. Pierre were crowded with people. After the explosion not one living being was seen on land. Only twenty-five of those on the *Roraima* out of sixty-eight were left after the first flash.[8]

The *Roraima*'s first officer, Ellery Scott, gave the following graphic account:

> About 8 o'clock loud rumbling noises were heard from the mountain overlooking the town, eruption taking place immediately, raining fire and ashes. . . . Soon . . . there came a terrible downpour of fire, like hot lead, falling over the ship

and followed immediately by a terrific wave which struck the ship on the port side, keeling her to starboard, flooding [the] ship, fore and aft, sweeping away both masts, funnel-backs and everything at once. . . . Shortly after, a downfall of red hot stones and mud, accompanied by total darkness, covered the ship. . . . I tried to assist those lying about the deck injured, some fearfully burnt. Captain Muggah came to me, scorched beyond recognition. He had ordered the only boat left to be lowered; but, being badly damaged, [it] could not be lowered from the davits.[9]

Only one ship escaped from St. Pierre—the *Roddam,* a steam-powered cargo vessel from England that had anchored in the roadstead only minutes before the fatal eruption. American authors Gordon Thomas and Max Morgan Witts in their book *The Day the World Ended,* a vivid account of the 1902 catastrophe, wrote that the *Roddam* was badly damaged, her deck a shambles, and most of her crew were dead or dying. But those who remained alive, though grievously injured, managed to get the ship under control. Captain Edward Freemen, badly burned himself and barely able to see because of injuries to his eyes, somehow guided the *Roddam* out of the roadstead and some 80 kilometers south to the neighboring island of St. Lucia, where at dusk he nosed her into the port of Castries.

A customs boat came out to meet the ravaged ship, and an officer called out "Where have you come from?" Cried Captain Freeman, "From the gates of Hell!"[10]

———

Mount Pelée continued to extrude its lava spine. A year after the eruption, it towered 300 meters above L'Étang Sec, piercing the air like a gigantic, malevolent finger offering a last gesture of Pelée's defiance, before slowly crumbling of its own weight.

St. Pierre was a dead city, in ruins after the pyroclastic flows of May 8. Some of the more substantial stone buildings, though heavily damaged, still stood—but on May 20 another devastating pyroclastic flow erupted from Mount Pelée. Following in the same path as the first, it burned or reduced to rubble whatever had managed to withstand the May 8 cataclysm. Looters soon appeared in the ravaged city, seeking money, jewels, and other valuables in the ruins. French marines eventually put a stop to the thievery, often shooting looters on sight, but bands of robbers terrorized the surrounding countryside for months.

Pelée erupted yet again in late August, this time destroying the village of Le Morne Rouge, a few kilometers northeast of St. Pierre. A short time before that eruption, the new governor of Martinique, Georges Lhuerre, had refused aid to refugees from Le Morne Rouge who had fled to Fort-de-France. Most, having no place else to go, returned home—and all were killed. Both Le Morne Rouge and St. Pierre have since been rebuilt, though St. Pierre today is a small town of minor importance. Ruins from the 1902 catastrophe remain plainly visible there.

The economy of Martinique was devastated by the May eruptions. Although much of the island escaped serious physical damage, virtually its entire commercial and cultural life was shattered with the destruction of St. Pierre. Moreover, ashfalls killed crops in the southern part of the island, and many farms and plantations were destroyed. And 30,000 people died in St. Pierre. The real tragedy of Mount Pelée is that the vast majority of those people surely would have survived if government policy and the lack of knowledge about the nature of volcanoes had not discouraged their evacuation.

The catastrophe had far-reaching political repercussions as well. The election scheduled for May 11 was never held, and the political ascendancy of the black and mixed-race citizens of Martinique was set back for decades.

MOUNT PELÉE AND THE PANAMA CANAL

In 1902, when Mount Pelée erupted, the U.S. Senate was preparing to vote on the location of a canal to connect the Atlantic and Pacific oceans. The world's maritime powers had long seen a need for a canal across Central America to avoid the long and dangerous voyage around Cape Horn, at the tip of South America. French engineers had failed in their attempt to dig a canal through the Isthmus of Panama, and the Americans were considering an alternative route through Nicaragua. The 1902 eruptions of La Soufrière on St. Vincent and Pelée on Martinique focused attention on the fact that Nicaragua, too, was subject to volcanic activity.

Politicians who favored the Panamanian route were quick to seize upon the Caribbean eruptions as evidence that building a canal in Nicaragua would be a risky venture. Indeed, that country had proudly featured a smoking volcano on its one-centavo postage stamp. A lobbyist for Panama obtained enough of the Nicaraguan stamps to send one to every member of the U.S. Senate, which subsequently, by eight votes, approved construction of the Panama Canal.

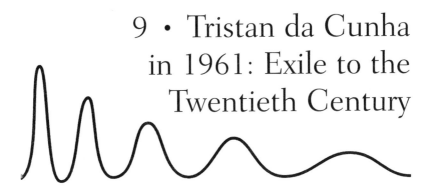

9 · Tristan da Cunha in 1961: Exile to the Twentieth Century

The death sentence was spoken on their way of life—
a way of life which was dearer to them than life itself.

Peter A. Munch, *Crisis in Utopia: The Ordeal of Tristan da Cunha*

THE ISLAND OF TRISTAN DA CUNHA is the most isolated of the earth's inhabited places. It is the top of a volcanic mountain that rises from the depths of the South Atlantic Ocean about 500 kilometers east of the axis of the Mid-Atlantic Ridge. At 37 degrees south latitude, Tristan is about as far south as Buenos Aires in Argentina and Cape Town, South Africa. Buenos Aires is almost 4,300 kilometers west, and Cape Town, more than 2,800 kilometers east (Figure 9-1, top).

Being the peak of a volcano, Tristan has an almost circular outline, with a diameter of only about 11 kilometers and a surface area of about 100 square kilometers. The top of the mountain is 2,060 meters above sea level. Its base lies on a flank of the Mid-Atlantic Ridge at a depth of 3,000 meters.

From a distance, the island appears bleak and barren. The only trees are stunted, wind-twisted evergreens. Narrow ravines radiate outward from the central peak, which more often than not is obscured by clouds. Ocean waves, whipped up by almost constant winds, crash against basalt cliffs as much as 600 meters high, white foam contrasting starkly with

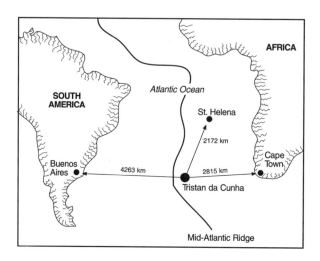

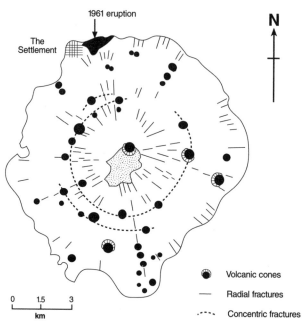

FIGURE 9-1. *Top:* The location of the island of Tristan da Cunha, which is far from any other inhabited land. *Bottom:* Tristan da Cunha is the top of a volcano. A number of volcanic cones developed on concentric as well as radial fractures.

the black rock. There is no natural harbor. Ships must anchor well offshore. Until a sheltered landing place was constructed for small craft in the late 1960s, the only access to the island was by rowing small boats through the surf onto one of the two rocky beaches not bordered by high cliffs or obstructed by dangerous rocks. The weather is often foul, with rain and high winds, though temperatures generally are mild.

Tristan da Cunha has been aptly described as "a gigantic ogreish mountain, wild, mysterious, and forbidding, rising out of the ocean where land is least expected."[1] Yet this remote, uninviting bit of land is home to some 300 people, almost all of them descendants of a few early-nineteenth-century settlers and shipwrecked sailors of various nationalities. The first settlers were mostly of British extraction, and the island eventually became a dependency of Great Britain. Sharing but seven family names, the inhabitants all live in Tristan's one town, officially named Edinburgh after a British duke who visited briefly in 1867, but called simply "the Settlement" by the islanders. The Settlement is a cluster of houses on a relatively flat plateau bordering the ocean at the northeast corner of the island.

Over the years the Tristan Islanders, or "Trist'ns," as they call themselves, evolved an anarchistic, almost utopian lifestyle. There was no government on the island, nobody "in charge." Everyone was the equal of everyone else. Each person went about his or her affairs independently, yet people went out of their way to help one another, and they cooperated in communal tasks such as fishing, gathering what food the island provided, building houses, and trading with passing ships. Well into the twentieth century there were no roads, no motor vehicles, virtually no modern conveniences. Tristan da Cunha was a relic of the nineteenth century.

And then, in 1961, the volcano erupted—not from the crater at its summit but from a new vent on a fissure less than 300 meters from the Settlement. Though the eruption was moderate, with a VEI (volcanic explosivity index) of only 2, there was no choice but to evacuate the island. The people were taken first to Cape Town, then to England, where they spent the better part of two unhappy years. Finally, in 1963, the British government reluctantly allowed them to abandon the putative benefits of the twentieth century, and they eagerly returned to their homes on Tristan. Their island, however, and they themselves, had changed irrevocably.

———————

Tristan da Cunha, two small islands nearby named Nightingale and Inaccessible, and associated submerged mountains are parts of a volcanic complex that resulted from emissions of molten rock, or magma, near the axis of the Mid-Atlantic Ridge. Crustal separation along the ridge has caused the complex to drift toward the northeast. The oldest lava exposed on the islands, on Nightingale, is about 18 million years old. On Inaccessible the lava is only about 6 million years old. Volcanic rocks on Tristan itself range from 700,000 to 3 million years old. The large mass of volcanic material beneath the island, however, suggests that Tristan's volcanism started much earlier.

The archipelago is believed to be the site of a hot spot—the surface manifestation of a plume of hot mantle material that has remained stationary for at least 120 million years. The plume's output of magma waxed and waned as crustal plates moved above it and drifted away during the opening of the South Atlantic Ocean (Figure 9-2). As the South American and African plates separated, they carried with them the accumulated hot-spot volcanic materials to form two major oceanic ridges, the northeast-trending Walvis Ridge and the northwest-trending Rio Grande Ridge. Westward movement of the Mid-Atlantic Ridge itself, relative to the mantle plume, is suggested by the fact that Tristan, an active volcano, is today almost 500 kilometers east of the zone of active volcanism where the plates diverge.

The directions of the hot-spot tracks shown in Figure 9-2 diverge markedly from the north-south trend of the Mid-Atlantic Ridge. The trend of the Walvis Ridge suggests that

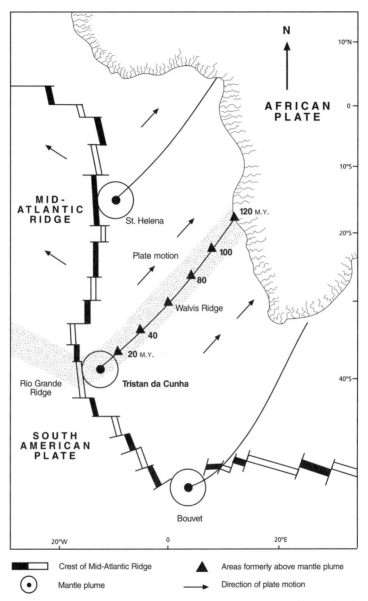

FIGURE 9-2. The tectonic setting of mantle plumes near the Mid-Atlantic Ridge. Also shown are paths of locations that were carried over the hot spots at different times in the past as the African plate moved northeastward. Numbers indicate millions of years (M.Y.). Adapted from O'Connor and le Roex, "South Atlantic Hotspot-Plume Systems," 356.

the African plate moved northeast while the oceanic crust in general was being formed by east-west spreading.

The implication is that the flow direction of mantle material beneath the plates differs from the direction of spreading along the Mid-Atlantic Ridge. The northwesterly and north-easterly hot-spot tracks imply that different layers of the mantle must have moved in different directions. While the lithosphere—the crust and solid upper mantle—spread east and west, the plastic asthenosphere beneath it moved in a northerly direction. The combined effect of those superimposed motions led to the present pattern of hot-spot tracks (shown in Figure 9-2).

The oldest volcanic rocks associated with the Tristan mantle plume are found in Brazil (the so-called Paraná flood basalts) and in East Africa (the Etendeka flood basalts). Those immense outpourings of lava occurred as the plume was breaking through the lithosphere 120 million years ago, when South America and Africa were joined. Separation of the South American and African plates was probably aided by the upwelling of several plumes in the vicinity of Tristan da Cunha, St. Helena, and the small island of Bouvet (see Figure 9-2).

The Tristan volcano is symmetrical and is characterized by a well-developed system of fractures radiating from the summit. These fractures are manifested on the surface as deep ravines, most containing streams that flow down to the sea like the spokes of a wheel. Many of the fractures were filled by magma that solidified into thin bodies of rock called dikes. Over some of the fractures, where magma reached the surface, usually at lower elevations, small volcanic cones developed (Figure 9-1, bottom). There are more than twenty such cones, most composed of fragmental, or pyroclastic, material.

The volcano's central vent has emitted both basaltic (silica-poor) and trachytic (silica-rich) lavas and pyroclastic materials. The more viscous trachytic lavas form the steeper upper slopes near the summit, while the more fluid basaltic flows underlie the less steeply inclined lower slopes.

Geological evidence suggests that the most recent volcanic event before 1961 occurred only 200 to 300 years ago at the Stony Hill cone, near the southern shore of the island. Since there is no historical record of that eruption, it probably occurred during one of the long intervals between early visits of Europeans to the island. The inhabitants of Tristan in 1961 were well aware of the volcanic origin of their island, but they believed the volcano to be extinct. Its renewed activity came as a cruel surprise.

———

Tristan da Cunha was discovered in 1506 by a Portuguese admiral named Tristão da Cunha. It was devoid of human habitation then and remained so for more than 280 years, though during the 1600s two or three Dutch expeditions from South Africa investigated its potential use as a naval base. Their reports were consistently negative, undoubtedly because they could find no natural harbor on the island and because of dangerous currents and unpredictable weather in the area. A century later, in 1760, a ship captained by a British officer named Gamaliel Nightingale visited the archipelago. The smaller of the two outlying islands now bears his name. And in 1778 a French captain, after futile efforts to find a landing place among towering black basalt cliffs circumscribing the third island, named it Inaccessible.

Thus the Tristan islands remained an undisturbed sanctuary for enormous numbers of fur seals, sea elephants, penguins, and sea birds of many kinds, and the surrounding waters were alive with whales and many varieties of fish. News of those riches inevitably filtered back to civilization, and during the late eighteenth and early nineteenth centuries small groups of men stayed on Tristan for brief periods, hunting fur seals for their skins, and sea elephants for the oil obtained by melting down their blubber.

Meanwhile that part of the South Atlantic Ocean had become well traveled by American and European sailors, whose wind-driven ships took advantage of the "roaring forties," the often fierce westerly winds that provided the fastest and most dependable route to the Cape of Good Hope and the ports of southern Asia and the East Indies. The high, conical profile of Tristan da Cunha was a familiar and welcome landmark in those trackless waters. Whaling ships and, during the Anglo-American war of 1812 to 1814, naval vessels, often anchored off Tristan to refit and take on water.

The first people to settle permanently on Tristan da Cunha were four men whose leader was one Jonathan Lambert of Salem, Massachusetts. They landed on the island in 1810, intending to profit from passing ship traffic by making the island a refitting station and by selling food and water. Lambert and two companions were lost at sea during a fishing trip, but the surviving member of the party, an Italian named Tomasso Corri, remained on Tristan for several years.

In 1816 Great Britain established a garrison on Tristan to forestall any attempt by the French to use the island as a base for rescuing Napoleon Bonaparte from his place of exile on St. Helena. The garrison was removed a few months later, the British apparently having realized that Tristan was too far from St. Helena to be of practical use to the French. One of its members, Corporal William Glass, received permission to remain behind with his wife, a fifteen-year-old girl of Dutch descent from South Africa. Two civilians attached to the garrison also chose to remain on the island.

The three men drew up a document stipulating that whatever goods they owned would be held in common, that profits made from sales of island products would be shared equally, that all work on the island would be shared, and that no one of them would be considered in any way superior to the others. That document, now in the British Museum in London, embodies the communal, noncompetitive spirit that has characterized the culture of Tristan da Cunha ever since.

Life on the isolated island cannot have been easy for the tiny colony. The men provided a living for themselves by bartering seal skins and sea-elephant oil to passing ships, and Corporal Glass and his young wife coped with the inevitable social tedium by producing sixteen children, eight boys and eight girls. Shipwrecked sailors increased the island's population from time to time. Some remained there and married Glass daughters. Others found wives when, in 1827, a ship captain agreed to seek volunteers on his next trip to St. Helena. He brought back five women of mixed heritage, all of whom found eager husbands. Most of the Glass sons joined the crews of American whaling ships, and some eventually settled in New London, Connecticut, a major whaling port at that time. Others brought wives back to Tristan. All the Trist'ns who were evacuated in 1961, after the volcano erupted, were descendants of those early settlers.

Tristan's barter economy originally flourished, in a modest way, from trading with the many sailing ships that passed the island. In addition to fishing, the islanders kept small flocks of sheep and a few cattle, and they grew potatoes. A fertile plateau about 3 kilometers southwest of the Settlement was divided into small plots, or "patches," separated by low stone walls. Each "potato patch" yielded perhaps 100 to 200 bushels a year. Frances Repetto, a leading member of the island community, was fond of saying "Every man on Tristan da Cunha can make his own living if he works on his potatoes, looks ahead, and saves for a rainy day."[2]

Whenever a ship hove into view, the islanders stopped whatever they might be doing and, if the weather was favorable, loaded their boats with trade goods—sacks of potatoes, seal skins, "elephant oil," or whatever else they might have to barter. They launched their boats through the surf from a beach near the Settlement and rowed out to intercept the ship, hoping to trade for worldly goods such as clothing, tools, lumber, hardware, and staple foodstuffs.

Though the islanders used lifeboats or other small craft salvaged from shipwrecks, they developed, over the years, a type of small, open boat uniquely suited to the conditions of their island. With no trees to cut down for lumber, they stretched stout canvas over a frame of whatever wood was available, including driftwood and the debris of shipwrecks.

The boats, up to 9 meters long, were light in weight yet strong and sturdy, and their framing was flexible enough to withstand pounding surf.

The Tristan Islanders were of necessity consummate seamen. Despite unpredictable weather, tricky near-shore currents, and the ever-present surf on their landing beaches, no one ever was lost in a boating accident until November 1885. Then, in a bitter tragedy, fifteen men disappeared in a lifeboat that had set out, with trade goods, for a ship that had been sighted several kilometers offshore. What happened to the boat remains a mystery. Islanders thought they saw it tied up to the ship, but the ship's captain, who saw the boat approaching, reported that it disappeared before reaching his vessel. In any event fifteen men, ten of them married, never returned home—a devastating blow to the small, close-knit community.

As the nineteenth century drew to a close, and sailing ships began giving way to steam-powered vessels, sea-lanes in the vicinity of Tristan da Cunha became less and less traveled. No longer dependent on the westerly winds of the roaring forties to reach the Cape of Good Hope, ships from Europe and North America took more direct routes. The whaling industry, moreover, began to decline after the discovery of oil in Pennsylvania in 1859, and fewer and fewer whalers called at Tristan. The remote island became more and more isolated, and its people were left to sustain themselves as best they could. During the late 1800s almost half of Tristan's population left the island, many in despair after the lifeboat disaster. By 1891 only sixty-three people remained.

Between 1886 and 1907 the British sent several envoys to Tristan in hopes of convincing the islanders to emigrate to South Africa. The feeling in London was that maintaining such a small colony on the remote island was more trouble than it was worth. A few of the remaining islanders agreed to go, but most of them steadfastly refused to leave their homes and their way of life for the uncertainties of the outside world.

Tristan's isolation ended during the 1940s, with World War II. The British stationed a naval garrison on the island, and some of the officers brought their families. Thus Tristan got its first school. A canteen built for the servicemen was open to islanders as well. They were able to buy items there with money earned in construction and maintenance work at the naval station and were thus introduced to a cash economy.

The naval station was abandoned after the war, but the South African government established a meteorological station on the island. Supply ships called regularly, and there was constant radio contact with Cape Town. Tristan da Cunha had entered the twentieth century.

Fishing companies in Cape Town became interested in exploiting the waters around Tristan. Those waters teemed with crayfish (also called crawfish or "rock lobsters"), which were in great demand for the American market. The Tristan da Cunha Development Company was formed in 1948 with exclusive rights to establish a fishing industry on the island. A canning factory was built near the Settlement in 1949 and later transformed into a freezing plant.

Two modern fishing boats plied the waters near the island. Tristan men were employed on the boats, and Tristan women, in the freezing plant. The industry brought, for the first time, a measure of prosperity to Tristan da Cunha—but it also brought an end to the islanders' cherished state of benevolent anarchy. In 1950 the British government appointed an administrator to oversee the relationship between the people of Tristan and the development company, a position that made the administrator, in effect, a governor.

Though the twentieth century had come to Tristan, the islanders did not accept it wholeheartedly. For most, their traditional values and lifestyle took precedence over regular jobs working for "the company." They still tended their potato patches, looked after their livestock, and helped one another with communal tasks. As one islander put it, "when you's workin' under a boss, . . . well, you's workin' to please somebody else, not to please yourself. But on Tristan I work to please myself."[3] The Trist'ns stubbornly kept to their old ways and maintained the values that had sustained them on their remote island for almost 150 years.

————

But in 1961 the volcano awakened. The first warning of a geological disturbance came on August 6, midwinter on Tristan. An earthquake with a magnitude estimated to have been 3 or 4 on the Richter scale shook the Settlement. Windows rattled and crockery fell from shelves—the first time such a thing had happened on the island. There were more tremors on the eighth and ninth, and on the tenth, six earthquakes followed one another in quick succession.

By the end of August the Settlement was being jolted at least two or three times a day. On Sunday, September 17, the strongest quake yet occurred while the islanders were in church for evensong. Peter Wheeler, the island's administrator, wrote, "suddenly the walls heaved, the floor trembled, and . . . the roof threatened to cave in."[4]

Wheeler sent urgent messages to Cape Town and London. The replies assured him there was no danger. Nevertheless, he organized scouting parties to determine the extent of the disturbances. A fishing boat carried one group to Nightingale, and other groups checked various parts of the island for evidence of seismic activity. The reports were negative. The earthquakes had been felt only in the immediate vicinity of the Settlement.

Early in October the seismic activity reached a climax. The shocks became more intense, probably not because of greater magnitude but because fissures were opening at increasingly shallower depths. Rocks and boulders rattled down from the high cliff behind the Settlement, and a landslide cut off the town's water supply.

Beneath the Settlement, intruding magma resulted in differential uplift, which produced cracks in the surface of the ground and in the walls of houses. On October 9 a rift about 240 meters long and perhaps 2 meters wide opened just east of the Settlement. A grazing sheep fell into a section about 3 meters deep. Bleating pitifully, the terrified animal tried to climb out, but the walls were too steep. Then the bottom of the fissure slowly began to rise, and as it approached ground level the sheep nonchalantly stepped out and continued grazing. This story, though possibly apocryphal, illustrates the opening of a fissure and how its bottom can rise as the magma pushes upward.

On the seaward side of the fissure, the ground began to rise and in less than two hours had formed a mound about 10 meters in diameter and 6 meters high. The mound continued growing, and the next day, October 10, it was 18 meters high and almost 50 meters across. A small volcano had been born (Figure 9-1, bottom). During the night it gave off an ominous red glow, reflected in clouds of white vapor that drifted skyward.

Fearing for the safety of the Settlement, the men of Tristan gathered in the town's meeting hall the afternoon of October 9 and devised a plan for immediate evacuation. Everyone, old and young alike, carrying little more than the clothes they wore, hiked the three kilometers to the potato patches. The fishing boats were hailed by radio and told to stand by for an evacuation to Nightingale the next morning. In "the patches" the islanders found shelter against bitter winter winds as best they could—in ditches, huddled in the lee of stone walls, even in unused barrels. A few unheated huts provided cold comfort for the old, the very young, and the infirm.

The plan was to take the islanders out to the fishing boats the next morning from a beach near the patches, but a rough surf made that plan unworkable. So on October 10 the weary people retraced their steps to the beach near the Settlement. Though later overrun by lava, the beach was still usable on the tenth. The roaring volcano provided a fiery and frightening backdrop as Tristan men in their canvas boats shuttled their friends and families to the waiting fishing vessels, which in turn carried the anxious refugees to temporary safety on Nightingale.

As it happened, a Dutch ocean liner was in the area. Contacted by radio, she detoured to Nightingale the next day and carried the islanders, with their few possessions, to Cape Town. The people of Tristan were given a warm welcome in Cape Town, but to their surprise and distress they did not stay there long. Authorities in London, still viewing Tristan da Cunha as a colonial liability, saw the eruption as "a godsend opportunity to get rid of a problem," as one bureaucrat put it.[5] Without consulting the islanders, the authorities had decided to permanently resettle the entire community in England. Thus only four days after they arrived in Cape Town, the hapless refugees were bundled onto another ship and taken to England.

By October 12 visiting British scientists had found that the mound near the Settlement had developed a series of craters, which were emitting fountains of red-hot clinkers. By the fourteenth the mound was more than 70 meters high and occupied more than 1.5 hectares. Lava began flowing seaward and by late October had buried the freezing plant.

The mound continued to grow slowly but steadily. By mid-December it was more than 150 meters high. About every ten minutes a crater at its summit puffed forth a white cloud, accompanied by a loud bang. Lava bombs shot almost 30 meters into the air. A second crater appeared and periodically emitted a sulfurous, yellow-gray cloud.

A second eruption center developed some distance southwest of the Settlement, in an area of wetlands. It discharged great volumes of hot mud as well as a yellow-brown vapor, which contrasted sharply with the white cloud arising from the original mound. That eruption ceased after two days, leaving a field of rich black soil.

By the January 5, 1962, the seaward-moving lava front was almost 1,200 meters wide. Activity continued throughout February and March, then gradually declined. Before the eruption ceased altogether, the lava had destroyed the freezing plant and overflowed the beaches used by the islanders for launching their boats.

Volcanic deposits from the eruption eventually affected almost 60 hectares of the Settlement area. Most of the Settlement itself, however, was spared. Lava flows covered 8 hectares, and pyroclastic materials, 32 hectares. Vegetation in the remainder of the area was badly damaged by sulfurous gases. Evidence from earlier eruptions elsewhere on the island suggests that the vegetation will not re-establish itself for several hundred years. All in all, the Tristan eruption was relatively minor and essentially nonviolent, but it significantly altered the one site that for almost 200 years had served the island's population as home.

———

The Tristan Islanders arrived in England on November 3, 1961. British authorities, acceding to the islanders' urgent request that they be allowed to remain together, sent them to temporary quarters in a disused army camp in Surrey, just south of London. There, at Pendell Camp, they were housed in a barracks with little privacy. It was a desolate place to begin a new life in an unfamiliar world.

Government agencies and charitable organizations provided financial aid and did what they could to help the Trist'ns adapt, and within a short time jobs were found for most of those who were able to work. The islanders were inoculated against the diseases of civilization, but during that first winter they suffered from colds, bronchial infections, and influenza, and four died of pneumonia. Worries about jobs, money, their homes on Tristan, and their future contributed to their malaise.

The search continued for a better place to resettle the people of Tristan, and in January 1962 they were taken to another government facility, Calshot, on the south coast of England near Southampton. There, in former quarters for married officers of the Royal Air Force, each family was provided with a furnished house. Essentially they now had their own community.

Problems remained, however. The islanders were concerned about health, finances, and returning to Tristan da Cunha, as well as crime, a facet of civilization that was entirely new to them. Confrontations with young hoodlums from nearby towns made them wary of strangers. They read about robberies and murders in British cities, and they grew fearful of leaving their homes.

As time passed, moreover, they became anxious about the government's intentions regarding their return to Tristan. The British had sent a scientific expedition to the island during the winter, and in April they brought back encouraging news—the eruption had ceased, the Settlement was largely undamaged, and the cattle had survived and were doing well—but government officials said nothing about returning the émigrés to their homes. "They can't keep us here?—Can they?" cried one woman in despair.[6] A month later, tired of waiting, the islanders composed a letter to the British Colonial Office stating their desire to go home and asking the officials to arrange transportation. The response was bureaucratic evasion. The islanders were put off with the excuse that any decision would have to be made by higher authority. Frustrated, they wrote another letter in July. That time, instead of asking permission to go, they said they *were* going, and if the Colonial Office could not help them, they would arrange transportation on their own. They emphasized the importance of returning to Tristan in time to get a new crop of potatoes planted before fall.

The government's response was to announce that a second scientific expedition would visit Tristan in the fall to evaluate the safety of the island. No decision would be made about the islanders' return until the expedition's report was in hand. The official position was that the volcano was still dangerous. In view of the eruption's cessation, the islanders failed to understand how Tristan could be more dangerous than England, with the crime and traffic accidents they read about every day.

The islanders held a meeting at Calshot and agreed that twelve men would return to Tristan immediately, as an advance party to prepare the island for the return of the rest. When the men applied for tickets at the Southampton office of the steamship line, however, they were refused passage. But then, possibly because of publicity in the news media about the islanders' plight, the Colonial Office reversed its position, called a meeting at Calshot, and announced that they *would* provide passage for the twelve men. The plan for a second scientific expedition was scrapped.

There was a caveat, however: a government official would accompany the advance party and would report to the Colonial Office on the safety of Tristan. No final decision would be made about resettlement until government officials had read the report. The bureaucracy was unwilling to rely on the judgment of the islanders themselves about the safety of their island. Nevertheless, the Trist'ns were overjoyed. They—at least some of them—were on their way home, and once there, they felt certain, nobody would have an easy time getting them to leave again, or keeping the others from joining them.

The advance party arrived on September 8. They found their village largely intact, with severe damage to only one house. The islanders' cattle not only had survived but also had produced a good many calves. The traditional boat-launching beaches near the Settlement were buried under lava flows, but another usable beach had been formed nearby from eruption debris. The lava was still warm and the new crater still smoking, but there was no doubt that the eruption had ended.

The official report was duly sent to the Colonial Office but was never made public. There was no word from government officials about returning the remaining islanders. Two months later, the Colonial Office announced that yet another advance party, to include an administrator, would go to Tristan to evaluate the situation there yet again.

To further complicate matters, the government had become convinced that many of the islanders wished to remain in England. The Trist'ns canvassed themselves and announced their readiness to return to their island. Government officials refused to accept that informal response and insisted upon a secret vote, which was taken in December. The result: 97 percent of the islanders voted to go home.

Rebuffed, and acting with bureaucratic deliberation, the British government did not send the second advance party to Tristan until April 1963. They chartered a Danish ship, the *Bornholm*, to carry the main group, but the *Bornholm* would not be available until October. Finally, on November 10, 1963—almost a year after the secret ballot and more than two years after the volcano had erupted—the *Bornholm* arrived at Tristan da Cunha with its happy passengers. The islanders cried with joy as they went ashore.

British news media, somewhat huffily, referred to the people of Tristan as having fled the twentieth century and rejected modern civilization. Author Peter Munch, in his book *Crisis in Utopia,* describes the media reaction this way: "After two long years as captive refugees in a highly industrialized society, with all its affluence and conveniences, the Tristan Islanders did indeed return to their own simple life before the very eyes of an amazed, dismayed, and somewhat insulted Western World. . . . It was as if our whole ethos and way of life had been put on trial, and had failed."[7]

The people of Tristan did not return to their island untouched by their two-year sojourn in England, of course. Most of them, especially the young, had adopted modern European dress, dietary preferences, and tastes in entertainment. They had become accustomed to modern appliances and furnishings and had brought them back to their homes. And their attitudes were different, especially regarding outsiders, whom they had learned not to trust implicitly. Most important, perhaps, they had achieved a greater sense of self-awareness, of their identity as Tristan Islanders. While retaining their cherished individuality, the Trist'ns had acquired a social cohesiveness.

A new fish-freezing plant was built, and a protected boat-landing was constructed at the new beach near the Settlement. Electricity was brought into the village, houses were modernized, and old paths were transformed into paved streets. Perhaps the Trist'ns had indeed rejected the twentieth century, but it nevertheless followed them home. A volcanic eruption brought it to this most isolated of the earth's inhabited places.

10 · Mount St. Helens in 1980: Catastrophe in the Cascades

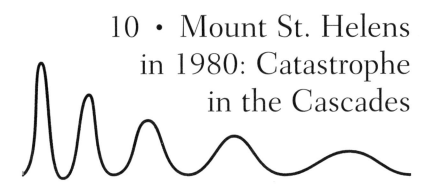

Those lovely white snow peaks, those rulers of mountains,
Who knows what secrets they keep?
But when they look like forever, just stop and remember
The Giants are only asleep.

Tom Shindler, "The Giants Are Only Asleep"

AFTER SLEEPING FOR 123 YEARS, Mount St. Helens, a volcano in the state of Washington, shook itself awake in 1980. A series of earthquakes beginning in mid-March, followed by minor eruptions of steam and volcanic ash, led up to a climactic blast on May 18. The top 400 meters of the mountain blew away, 500 square kilometers of the Gifford Pinchot National Forest became a gray, ash-covered wasteland, and fifty-seven people died. Great clouds of volcanic ash darkened the region's skies, and dust from the eruption drifted around the Northern Hemisphere. It was the first volcanic eruption in the contiguous United States since California's Lassen Peak was active from 1914 to 1917.

As impressive—even terrifying—as it was, the 1980 eruption of Mount St. Helens was small in comparison with many other volcanic eruptions the world has known. We include Mount St. Helens here not because of any special geological or historical importance, but because it provides a recent, intensely studied, and highly publicized example of the ways in which a single geological event can affect so many aspects of human life. The 1980 eruption was the most photographed and the most thoroughly documented volcanic event in history.

Mount St. Helens is one of fifteen volcanoes in the Cascade Range, which extends from northern California into Canada's British Columbia. The youngest of the Cascade volcanoes, St. Helens first erupted only 40,000 to 50,000 years ago—and the peak that disintegrated in 1980 was only about 2,500 years old.

For the past 4,000 years the Cascades have averaged one or two eruptions per century. During that time Mount St. Helens has been the most active volcano, though the periods between eruptions have been irregular. Recurrence intervals have ranged from about 40 to 140 years, although at times the volcano seems to have been inactive for hundreds of years. There is no evidence of volcanism, for example, between 1610 and 1800. The 1800 eruption was followed by fifty-seven years of intermittent activity.

It was during that period that Mount St. Helens acquired the symmetrically beautiful shape, rising 2,951 meters above sea level, for which it was noted until May 18, 1980. On that day the graceful cone blew apart, leaving a grotesque, lopsided crater where the mountaintop had been. The yawning depression is 625 meters deep, 2 kilometers wide, and 2.7 kilometers long. It is open to the north, for that side of the mountain broke loose in a colossal, earthquake-triggered avalanche that initiated the May 18 eruption.

The Cascade mountains represent a volcanic arc created where the small Juan de Fuca tectonic plate is moving eastward and sliding, or subsiding, beneath North America (Figure 10-1). The Juan de Fuca plate originated as upwelling molten rock, or magma, through rifts in the Juan de Fuca Ridge, the remaining northern segment of a larger oceanic feature called the Farallon Ridge. Most of the Farallon Ridge has been overridden by the westward-drifting North American plate.

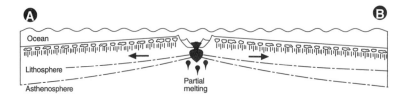

A ⓐ ▬ ⓑ B

Ocean

Lithosphere

Asthenosphere

Partial melting

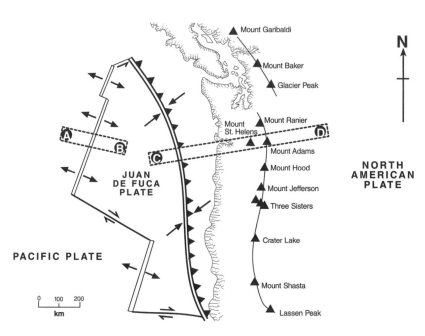

▲ Mount Garibaldi

▲ Mount Baker

▲ Glacier Peak

Mount Ranier

Mount St. Helens

ⓐ ⓑ ⓒ ⓓ

▲ Mount Adams

▲ Mount Hood

▲ Mount Jefferson

▲ Three Sisters

JUAN DE FUCA PLATE

NORTH AMERICAN PLATE

N

▲ Crater Lake

PACIFIC PLATE

▲ Mount Shasta

▲ Lassen Peak

0   100   200

km

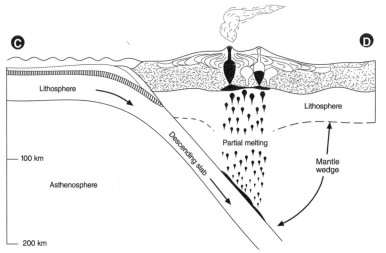

C ⓒ ⓓ D

Lithosphere

Lithosphere

100 km

Descending slab

Partial melting

Mantle wedge

Asthenosphere

200 km

At a depth of about 100 kilometers, the rock that makes up the subsiding Juan de Fuca plate is heated to such a degree that hot fluids, mostly water, are driven out. Penetrating into the overlying wedge of mantle rock, these hot fluids cause chemical interactions that lower melting temperatures in the wedge. Thus some of the mantle material melts, forming blobs of magma that rise through fractures in the crust and accumulate in chambers beneath the volcanic peaks of the Cascade Range (see Figure 10-1).

The size of the ensuing eruptions has varied greatly. About 6,000 years ago one of the Cascade volcanoes, known as Mount Mazama, blew up in a stupendous eruption that deposited a thick blanket of volcanic ash over all of what is now the northwestern United States. Today Mazama's caldera, in southwestern Oregon, is occupied by Crater Lake, almost 10 kilometers across and 610 meters deep. There is no geological reason why another of the Cascades volcanoes, including Mount St. Helens, could not erupt in a similar cataclysm at any time in the future.

Mount St. Helens was named in 1792 by Commander George Vancouver of the British ship *Discovery* while he was surveying the Pacific coast of North America. At that time Spain claimed most of what is now the western United States, and Vancouver named the volcano after Britain's ambassador to Spain, Alleyne Fitzherbert, also known as Baron St. Helens.

Native American tribes had names for Mount St. Helens and neighboring peaks that reflected prehistoric myths about their volcanic behavior. In one myth the beautiful, symmetrical peak Vancouver named St. Helens was once a comely

FIGURE 10-1. *Opposite:* The configuration of the Juan de Fuca plate, showing the zone of crustal spreading where new lithosphere is generated (section A-B) and the collision zone with the North American plate, along which the Juan de Fuca plate is being subducted (section C-D). To the east is the Cascades volcanic arc. Also shown in C-D is the generation of magma in the mantle wedge above the descending slab, and its rise and accumulation in magma chambers beneath Mount St. Helens.

maiden called Loo-wit. Two sons of the Great Spirit, Wyeast and Pahto, fell in love with Loo-wit, but the coy maiden would not decide between them. Wyeast and Pahto fought bitterly over her. The Great Spirit, angered at their behavior, turned all three into volcanoes. In their continuing fury, Wyeast (today's Mount Hood) and Pahto (Mount Adams) breathed clouds of hot ash, threw red-hot rocks at each other, and destroyed villages and forests with streams of liquid fire or buried them in mudflows. Meanwhile Loo-wit (Mount St. Helens), in her pristine beauty, presumably looked on quietly.

By the mid-1900s St. Helens, only 80 kilometers from Portland, Oregon, and not much farther from major population centers in Washington, had become the focus of a magnificent wilderness recreational area noted for hunting and fishing, hiking, camping, and skiing. Picturesque Spirit Lake, nestled in a valley on the north flank of the mountain, was a popular resort. Its poetic name originated in Native American legends about moaning noises and the mysterious disappearance of canoeists on the lake.

In an instant the 1980 eruption forever changed Spirit Lake and all the countryside for a great distance around. Beautiful coniferous forests, sparkling lakes, and clear-running streams gave way to frightful devastation. The area remains a major tourist attraction, but today the attraction is the havoc wreaked by titanic forces of nature—and also the natural healing of biological regeneration as new plants take root and wildlife returns.

As early as March 20 an earthquake of magnitude 4.1, caused by strain release along a fault beneath the mountain, alerted geologists to the possibility of an eruption. Scientists of the U.S. Geological Survey (USGS) immediately began installing gravity meters and tiltmeters to detect uplift and deformation of the ground. They placed automatic time-lapse cameras around the mountain to record any visible activity and installed gas sensors to detect any emissions of sulfur dioxide, which would indicate magma rising within the volcano's conduit. They also installed a network of seismographs to detect earthquake activity in the vicinity. The instruments recorded many low-magnitude quakes caused by the fracturing of rocks as magma inexorably rose through the conduit. They also recorded more or less continuous low-frequency vibrations, or harmonic tremors akin to a humming noise, probably caused by the separation of gases from the magma.

The shaking and humming continued, and geologists and Forest Service personnel began warning people to stay away from the area. Many, however, were not ready to accept an element of danger in the natural beauty of their surroundings. The notion of a harmful volcanic eruption in that serene, forested landscape hardly seemed credible. But just a few years earlier, in 1974, Dwight Crandell and Donal Mullineaux of the USGS had published a paper on volcanic hazards in the Cascade Range. Their warning was explicit: "The Cascade Range volcanoes have been so peaceful during the present century that there has been virtually no concern for potential volcanic hazards. As a result, dams and reservoirs have been built in valleys which have been repeatedly affected by large mudflows . . . or lava flows in the very recent geologic past."[1] The authors estimated that an eruption of Mount St. Helens could endanger 40,000 people. Those and similar reports, however, were written for geologists and did not receive wide publicity.

In 1978 Crandell and Mullineaux published another paper, which did reach a wide audience. Prophetically, they wrote, "In the future, Mount St. Helens probably will erupt violently and intermittently just as it has in the recent geologic past, and these future eruptions will affect human life and health, property, agriculture, and general economic welfare over a broad area. . . . an eruption is . . . likely to occur within the next hundred years, and perhaps even before the end of this century."[2] Their predictions, however, did not necessarily agree with the public's perceptions of volcanic risk. Further, the issue was clouded by misunderstandings on the part of the news media, government officials, and even some geologists.

A large part of the problem was that no one had much experience in predicting volcanic activity, and most scientists

were reluctant to make firm predictions because a false pre-
diction almost certainly would undermine public confidence.
Worse, it could lead authorities to order a costly evacuation
that later would prove unnecessary.

Moreover, many scientists were inexperienced in the art of
clearly briefing members of the media and local government
officials who had no background in the geological sciences. In
addition there was much confusion as to the specific respon-
sibilities of the various government agencies, and there was a
lack of interagency cooperation. As a result, the early warnings were largely ignored by a skeptical public.

By late March, however, most officials had been convinced
of the potential danger, both by ominous reports from the
geologists monitoring the volcano and by a widely dissemi-
nated hazard map that Crandell and Mullineaux had included
in their 1978 paper. On March 26 federal, state, and local government officials met with geologists and Forest Service personnel to plan emergency procedures. The USGS and the Forest
Service established a coordinating center in Vancouver, Washington. Hundreds of people, many under protest, were evacuated from mountain homes, campgrounds, and logging camps
near the mountain.

On March 27 molten magma rising within the volcano's
conduit contacted groundwater in the surrounding rock, sending a plume of steam and ash about 3 kilometers into the sky.
The eruption produced a small crater in the mountaintop and
scattered ash across pristine snow and ice near the summit.
The event became a media bonanza. Stories appeared in news
magazines under such flamboyant titles as "Beginning of the
End" and "Life in the Shadow of a Killer Volcano." Inevitably
thousands of sightseers descended upon the area, ignoring
warnings to stay away. Anticipating a larger eruption, news
reporters and television crews came from near and far. Air
traffic above the mountain became so heavy that the Federal
Aviation Administration had to restrict overflights.

Roads leading into the area were jammed with vehicles.
A carnival atmosphere prevailed as vendors hawked bumper

stickers, pennants, hats, Frisbees, packages of volcanic ash, and T-shirts emblazoned with silly slogans such as "I Lava Volcano." Restaurants sold "Volcano Burgers" and "Inferno-dogs." Roadblocks were installed to keep people from getting dangerously near the volcano. Many of the roadblocks were not staffed and were removed or bypassed by those who felt they had important business in the area or, skeptical of the danger, were eager to get a close-up view of a rare natural spectacle.

Steam and ash eruptions on March 28 and 29 created a second crater at the volcano's summit. Steam eruptions and earthquakes continued, the quakes becoming shallower as magma rose higher in the volcano's conduit. On April 3 the two craters coalesced to form a single depression about 500 meters across and 260 meters deep. That day the governor of Washington declared a state of emergency, and National Guard troops were called in to protect the roadblocks.

On April 12 aerial photographs revealed that the north flank of the mountain had begun to bulge, pushed out by rising magma. The bulge already had risen some 90 meters. Like a gigantic blister, it continued to grow at the rate of a meter and a half a day. By April 29 it was a kilometer and a half long and almost a kilometer wide. On April 30 the governor defined a "red zone" around Mount St. Helens, within which public access was denied to all but authorized personnel.

On May 1, to better observe the bulge and watch for any mudflows or avalanches, the USGS established an observation post on a ridge 10 kilometers northwest of the volcano. The post was manned by Harry Glicken, a recent graduate of the University of California, Santa Barbara. Glicken was working as a field assistant to David Johnston, a USGS geologist. The observation post was in radio contact with the Vancouver coordinating center. As it happened, Glicken had made an appointment months before to be in California on May 18 to discuss graduate work at the university with his faculty advisor. So that Glicken could fly to California, Johnston replaced him on May 17.

Mount St. Helens exploded at 8:32 the next morning. Within seconds the observation post, thought to be safely distant from the volcano, was swept to oblivion by a monstrous blast fiery gases, volcanic ash, and rock debris—a pyroclastic flow—that shot laterally from what had been the bulge on the north side of the mountain. The blast moved at speeds estimated to have approached 500 kilometers an hour. David Johnston's last words, shouted into his radio microphone, were, "Vancouver! Vancouver! This is it!"

That climactic eruption was preceded by an earthquake of magnitude 5.1, which caused the bulging, unstable north side of the volcano to slide away in three great blocks. The landslide blocks broke up and coalesced into an enormous avalanche of rock, glacial ice, snow, and other debris that roared down the mountainside at more than 200 kilometers an hour. It was among the largest landslide-debris avalanches ever recorded. The avalanche relieved pressure on magma beneath the bulge, releasing the pyroclastic flow that killed David Johnston.* Immediately afterward a vertical eruption column shot skyward, carrying an estimated one-half billion tons of ash and other debris to a height of some 25 kilometers (Figure 10-2, bottom). That phase of the eruption, which continued for nine hours, had an estimated VEI (volcanic explosivity index) of about 5, considered "very large" by volcanologists.

Meanwhile the pyroclastic flow overtook the avalanche, sped over it, and swept across four high ridges—the one on which the USGS observation post was located has since been named Johnston Ridge—and devastated an area of more than 500 square kilometers (Figure 10-2, top). Temperatures within the flow, later estimated from charred wood and scorched vehicles, must have been as high as 300 degrees Celsius. Within that area, trees by the millions—some as much as 50 meters

---

*Harry Glicken wrote his doctoral dissertation on volcanic debris avalanches like the one on Mount St. Helens that led to the death of his colleague David Johnston. Tragically Glicken, too, was killed by a pyroclastic flow in June 1991 while investigating an eruption of a volcano named Unzen, near Nagasaki, on the island of Kyushu in Japan.

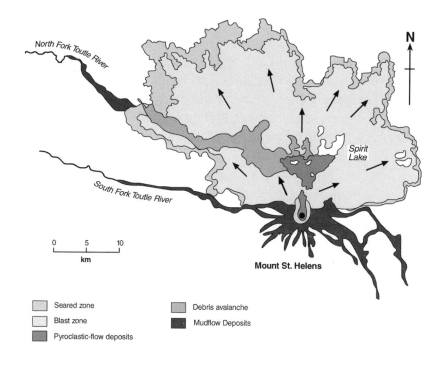

North Fork Toutle River

South Fork Toutle River

Spirit Lake

0   5   10
km

Mount St. Helens

▦ Seared zone
▢ Blast zone
▨ Pyroclastic-flow deposits

▨ Debris avalanche
■ Mudflow Deposits

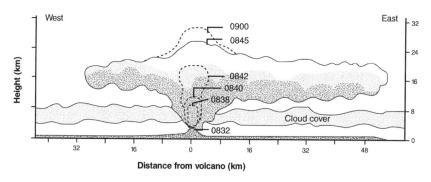

West   East

0900
0845

0842
0840
0838

Cloud cover

0832

Height (km)

Distance from volcano (km)

FIGURE 10-2. *Top:* The area devastated by the 1980 eruption of Mount St. Helens, and the distribution of pyroclastic flows, pyroclastic debris, and mud flows after the eruption. *Bottom:* The change in the vertical and horizontal extent of the ash cloud during the first half hour after the eruption. From Tilling, *Eruptions of Mount St. Helens.*

tall and 5 meters in diameter—fell in rows like gigantic matchsticks. Their snapped-off trunks lay parallel to the various directions taken by the pyroclastic flow as it swirled around hills and eddied along the contours of valleys. This "blowdown zone" has remained an area of awesome destruction. It originally contained an estimated 3.7 billion board feet of lumber, a considerable quantity of which was salvaged after the eruption. Virtually all wildlife within the blowdown zone was destroyed as well.

The avalanche, which originally had preceded the pyroclastic flow, was overpowering in its mass and volume. It comprised more than 2 cubic kilometers of broken rock, some pieces the size of large buildings, as well as chunks of ice, enormous quantities of volcanic ash and dirt, and millions of tree trunks. That juggernaut, moments after the flow, slammed into Spirit Lake, which was 8 kilometers from the summit of Mount St. Helens. The avalanche deposited as much as 90 meters of debris on the lake bottom and generated waves that washed up shoreline mountainsides as high as 75 meters. Avalanche material subsequently dammed the lake's outlet, raising the water level by about 60 meters and doubling the size of the lake. Whatever may have remained of an inn and several resort cottages at the south end of the lake lay deep beneath the avalanche and the raised lake waters, upon which floated countless tree trunks and an odious agglomeration of other debris. The bottom of the lake today is at a higher elevation than the surface of the water before the eruption.

Most of the avalanche surged west of Spirit Lake and hurtled almost 25 kilometers down the valley of the North Fork Toutle River, which drains the lake and flows northwestward (Figure 10-2, top). A hummocky mass of rocks and chunks of ice, intermixed with finer debris, filled the river valley to an average depth of almost 50 meters. In some places it was more than 180 meters deep.

More ground-hugging pyroclastic flows separated from the eruption column above the mountain or jetted from the volcano itself. They deposited a final layer of gray ash and pumice on an already ravaged landscape. Where there had been evergreen forests, sparkling lakes, and clear-running streams, there was now a bleak, barren, otherworldly desert from which all color—and virtually all life—had been taken away.

The pyroclastic flows melted snow and ice on the mountain. More meltwater—billions of liters of it—came from avalanche ice covered with hot volcanic ash. All that water, along with groundwater from the mountain itself, formed mud-carrying streams that gathered more mud and ash and became colossal mudflows. Inexorably the mudflows, moving at about 40 kilometers an hour with the consistency of wet cement, poured down the river valleys flanking Mount St. Helens (Figure 10-2, top). They deposited so much mud and debris in the upper reaches of the North Fork Toutle River valley that its floor was raised as much as 180 meters. New lakes and ponds formed where the deposits dammed tributary valleys.

Some 2,000 people were hastily evacuated, some by helicopter, from the path of the advancing mud. Gaining speed as the mud was thinned by river water, the onslaught—carrying millions of logs like battering rams—destroyed more than 200 houses in the valley, wrecked bridges, and overturned bulldozers and heavy-duty logging trucks. Finally a flood of water, heavily laden with mud, poured into the Cowlitz River, which in turn dumped enormous quantities of sediment into the Columbia River, clogging its shipping channel. All told, the floods dumped an estimated 73 million cubic meters of sediment into the Cowlitz and Columbia rivers. Upstream in the Columbia, more than twenty oceangoing ships were trapped at Portland and Vancouver until the U.S. Army Corps of Engineers deepened the channel with dredges working around the clock.

Post-eruption studies of the North and South Fork Toutle rivers showed that their load of suspended sediment was 500 times greater than before the eruption.[3] Even after twenty years, the sediment load remained 100 times pre-eruption amounts, showing that the aftereffects of geological instability can last for decades.

―――――――

The sound of Mount St. Helens' explosion carried at least 800 kilometers. To people in western Montana, two states downwind, the blast sounded like artillery fire. Yet people near the volcano heard nothing. They became aware of the eruption only when they felt the force of the blast, heard trees falling, or saw the eruption cloud fill the sky. In Portland, only 80 kilometers from the volcano, people had to hear about the event on their radios. The main reason apparently is that the sound waves rose almost vertically, then bounced back to earth at an angle from a layer of warm air about 5,000 meters up in the stratosphere. The rebounding waves left a ring of silence between the crater and the area where they returned to earth about 130 kilometers away. In addition, sound waves near the ground undoubtedly were absorbed and scattered by clouds of volcanic ash, much as sound is muffled by a winter snowfall. Because the region is well populated, the zone of silence could be mapped in some detail and has helped to explain similar phenomena during eruptions elsewhere in the world.

The eruption would have killed many more people than it did if not for the efforts of civil authorities, persuaded by geologists monitoring the volcano, to restrict access to the area. Surely, without those restrictions, thousands of sightseers, property owners, and others would have been killed. As it was, only fifty-seven people are known to have died, most by inhaling volcanic ash and suffocating. Providentially the eruption occurred on a Sunday, when very few loggers, who had permission to be in the forests, were at work. Washington National Guard helicopters rescued more than a hundred survivors within hours of the eruption, and rescue parties on the ground located almost a hundred more within a few days. A fortunate few apparently survived because eddies in the pyroclastic flows spared them the full effects of the blasts.

Farther from the volcano, ash from the vertical eruption plume, carried eastward by high-altitude winds, fell to earth like heavy, black snow across much of eastern Washington and into Idaho and even western Montana. High in the stratosphere, clouds of volcanic dust drifted across southern Canada and much of the United States. They reached the east coast in three days and circled the globe within seventeen days.

The volume of ash produced by Mount St. Helens was small compared with that produced by eruptions of volcanoes such as Tambora in 1815 and Krakatau in 1883. Less than 8 centimeters fell in eastern Washington and Idaho, but in today's technological society the ash caused ruinous problems. Airborne ash short-circuited electrical transformers, causing power outages. Clogged air filters disabled motor vehicles. Ambulances, police cars, and other emergency vehicles became inoperable. Thousands of stranded motorists had to find shelter in motels, public buildings, even private homes. Buses and trains stopped running, and airplanes were grounded. Highways were closed, and many businesses had to shut their doors. Moreover, the ashfall locally turned day into night, and breathing was so difficult that many people had to wear face masks to filter out the dust.

A narrow path of greatest ashfall thickness extended east-northeastward across Washington and into Idaho. Directly in that path was the town of Ritzville, 300 kilometers from the volcano. There, almost 8 centimeters of ash covered the ground. The city of Yakima, 137 kilometers northeast of Mount St. Helens, was much closer to the volcano but received only a little more than a centimeter of ash because strong high-altitude winds split the eruption cloud. Nevertheless work crews in Yakima removed 600,000 tons of ash from roofs, streets, sidewalks, and parking lots—a job that took ten weeks and cost more than 2 million dollars.

The area of devastation around Mount St. Helens has provided a natural laboratory for the study of volcanic processes and their ecological effects, including the renewal of life after such a catastrophe. The most critical agents of biological regeneration were not colonizing animals and plants from outside the area of devastation, as scientists had expected, but organisms that survived the eruption. Surprisingly, some small animals lived through the holocaust. Fish survived in an ice-covered lake. Frogs and salamanders that had burrowed into mud or were under water during the blast survived it, as did some crayfish. And tracks later found in the cooled ash indicated that a few beavers and gophers survived in their underground burrows. Moles and ants survived as well, as did roots and bulbs of some wildflowers, and even small trees and shrubs that had been buried in snow. Scientists coined the term *biological legacies* for these scattered survivors, so vitally important to the process of renewal.

Even twenty years later, however, at the time of this writing, the area remains in a state of ecological disequilibrium. Animals that used to live high up on Mount St. Helens are now found at much lower elevations, and animals that are rare or endangered elsewhere now flourish. Biologists believe it may take hundreds of years for the ecology of the area to stabilize and for a mature forest to return.

———

The destruction, economic hardships, and social dislocations resulting from the eruption of Mount St. Helens inevitably led to emotional problems for many people. Many of those living nearby had to leave their homes, some permanently. Others chose to move out of the area altogether. A forced move, however, is a cruel choice for most families. Most breadwinners have local jobs, and families typically have support networks of friends and relatives. Giving up jobs and severing such relationships can be traumatic. Moreover, people who own homes may not have the financial resources to move because they cannot find buyers for their houses. Many people who remained in the area suffered financially because of lost business or the costs of repair and cleanup.

The continuing threat of future volcanic activity raised a number of psychological issues. For example, some people made a scapegoat of Mount St. Helens, blaming the volcano for personal problems unrelated to the eruption. In 1982 the Federal Emergency Management Agency published results of a study undertaken by four investigators from the University of Minnesota.[4] The study was based on interviews with people living in three metropolitan areas—Longview and Kelso, 56 kilometers west of the volcano; Yakima, 137 kilometers northeast; and Pullman, 402 kilometers east.

People who lived in Yakima (pop. 52,000) had suffered the most because, although the city received only a little over a centimeter of volcanic ash, it was paralyzed for days. The citizens of Pullman (pop. 23,500) were affected the least and suffered little stress. Those in Longview (pop. 31,500) and Kelso (11,800) also suffered little, at first, because they were west of the volcano, and prevailing winds blew most of the eruption products eastward. Longview and Kelso lie at the junction of the Columbia and Cowlitz rivers, however, and only about 20 kilometers north, the Toutle River flows into the Cowlitz. The small town of Toutle, a short distance up the Toutle River, had to be evacuated after the eruption because of devastating mudflows.

The evacuation led to apprehension downstream in Longview and Kelso. In October 1980, six months after the eruption, federal officials warned the residents of those towns that thousands of households probably would have to be evacuated if watersheds upstream, on denuded mountainsides, received normal amounts of rain and snow that winter. It was feared that thick deposits of volcanic ash on mountain slopes could liquefy to form massive mudflows that would sweep down the valley of the Cowlitz.

In December stress levels among the residents of Longview and Kelso shot up. Heavy snowstorms and intermittent rains threatened to cause mudflows, and there were rumors of imminent evacuation. Fortunately, precipitation for the most part was abnormally low during the following months, and large mudflows did not materialize. The study showed that anxiety about possible aftereffects caused almost as much stress as the volcanic eruption itself.

Another study, conducted in 1984 by P. R. and J. R. Adams of Utah State University, found that Mount St. Helens could be held accountable for an 18 percent increase in the death rate, a 21 percent increase in emergency-room visits, and a 200 percent increase in stress-aggravated illness in the seven months following the eruption. Further, Adams and Adams found evidence for a 37 percent increase in aggressive behavior, a 45 percent increase in domestic violence, and a 235 percent increase in mental illness.[5]

———————

Another aftereffect of Mount St. Helens' eruption was a surge in letters to the editors of local newspapers concerning the event. And there was a veritable avalanche of publications about the eruption. In 1984 Scarecrow Press in Metuchen, New Jersey, published *Mount St. Helens: An Annotated Bibliography*, compiled by Caroline Harnly and David Tyckoson. After querying libraries throughout the United States, Harnly and Tyckoson catalogued 1,738 works published in the thirty-four months from March 1980, when earthquakes first signaled the pending eruption, to December 1982. The author of one of the cited articles predicted, perhaps a bit hyperbolically, "There will be a mountain of scientific papers" and "by the time they've finished years hence, they will have produced enough paper to fill the crater."[6]

Harnly and Tyckoson's bibliography divides publications into fourteen categories, as shown in Table 10-1. The categories are necessarily general, and inevitably there is some overlap among them. The publications include reports in scientific and technical journals, articles in magazines intended for the general reader, and books both scientific and popular, as well as maps and Ph.D. dissertations.

Much more has been written about Mount St. Helens, of course, since 1982. Scientists continue to study the volcano and publish the results of their work. Journalists continue to file news reports about the mountain. Authors continue to write

TABLE 10-1. Publications about Mount St. Helens, 1980–1982

NO.	CATEGORY	TOPICS INCLUDED
407	Geological studies	Volcanology, seismology, hydrology, etc.
306	General information	Nonspecific articles and news items
171	Industry, engineerinng	Effects on industry and infrastructure
131	Before March 20, 1980	Past eruptions, state of mountain before March 20
119	Biology, environment	Plants, animals, ecosystem
113	Atmosphere, weather	Gases, ash in atmosphere, effects on weather
104	Chemistry, physics	Chemical, physical properties of volcano
99	Medicine, health	Effects on human health
92	Business, economy	Economic, legal, commercial implications
81	Society, culture	Social and cultural aspects of the 1980 eruption
62	Agriculture	Agricultural practices, equipment losses
36	Books	Books written about the 1980 eruption
10	Maps	Publications consisting solely of maps
7	Dissertations	Ph.D. dissertations dealing with Mt. St. Helens

SOURCE: Harnly and Tyckoson, *Mount St. Helens.*

books about it. And semblances of the catastrophic 1980 eruption, whether realistic or not, have appeared in motion pictures such as *Dante's Peak,* released in 1997.

Inevitably the paroxysm of Mount St. Helens also inspired poets, artists, and photographers, eager to communicate thoughts and emotions generated by the event—whether grief because of the human tragedy, elation at the magnificence of a wondrous spectacle, or vivid descriptions of a dramatic act of nature. In 1981 the National Speleological Society published a poem titled "Remember Spirit Lake." Signed only "by Cricket," the poem is remarkable for its concise and beautiful word picture of the geological sequence of events on that fateful day:

Mountain sleeping
Woodlands welcome
Bird upon the wing

Morning drifting
Through the meadows
Gentle mists of spring

Mountain dreaming
Endless visage
Silken clouds on blue

Sunlight dancing
On the waters
Pebbles shining through

Mountain stirring
Woodlands tremble
Earth beneath my feet

Still the drifting
Dreaming dancing
Springtime air is sweet

Mountain waking
Quaking breaking
Open to the sky

Lightning ripping
Through the meadows
Ashes blowing high

Heavens boiling
Hillsides burning
Rivers turn to steam

Forests hurtling
Valleys choking
Lakes and spirits scream

Mountain building
Birthing bleeding
Mighty act of pain

I am watching
Wondrous weeping
Till the mountain sleeps again[7]

In a similar vein a song entitled "The Giants Are Only Asleep," composed by a musician named Tom Shindler, poetically describes the eruption and its devastation. It concludes with the following verses:

Is it really any wonder when you hear the mountains
    thunder
Just like they've done so many times before?
Funny, how so many folks were taken by surprise,
Like volcanoes never happen anymore.

Those lovely white snow peaks, those rulers of
    mountains,
Who knows what secrets they keep?
But when they look like forever, just stop and remember
The Giants are only asleep.[8]

Painters and photographers, amateur and professional alike, flocked to Mount St. Helens to capture the magnificence of the eruption and the frightful devastation of its aftermath. Tragically, some died in the pursuit of their craft. Others produced unforgettable pictures that dramatically reveal nature's awesome power.

————

After the 1980 eruption, viscous magma continued to well up in the throat of Mount St. Helens, forming a sequence of volcanic domes within the crater. The domes were repeatedly shattered by new eruptions, some sending clouds of debris more than 15,000 meters into the air. By 1986, however, the volcano appeared to have returned to a dormant state. Figure 10-3, from a photograph taken in 1994, dramatically shows a crater dome, the wide gap in the northeast side of the crater, and Spirit

FIGURE 10-3. The gaping crater of Mount St. Helens in the summer of 1994, with Spirit Lake in the background. Note the post-eruptive dome in the middle of the crater. Photograph by Pierre Rollini, used with permission.

Lake nestled among mountain flanks that show early signs of renewed greenery. On the horizon looms Mount Rainier, reminiscent of Mount St. Helens as it looked before the catastrophe of 1980.

Since then volcanologists have made considerable progress in attempting to predict eruptions in the short term. There is still no way, however, to predict the exact time, the size, or the duration of an eruption; and long-range predictions remain beyond the capabilities of today's science. No one knows how long Mount St. Helens will remain dormant or when the next catastrophic eruption will occur. Meanwhile geologists can only keep analyzing the volcano's activity—gathering data, watching, and waiting.

From wellspring of Native American legends, to majestic centerpiece of a forested area of beauty and recreation, to its devastating 1980 explosion and today's ugly stump of a mountain, St. Helens continues to focus our attention on the many faces of volcanism. Let the remnant of that once-majestic snow peak in the scenic Cascade Range continue to remind us that "the Giants are only asleep."

• Afterword

IN THIS BOOK we have tried to bring volcanoes to life, so to speak. We have tried to give them a human dimension, showing, with the aid of the "vibrating string" metaphor, how their aftereffects can resonate in human affairs for years, decades, centuries, or millennia. At the same time we have tried to explain, in terms of plate tectonics, why volcanism exists and how selected volcanoes came to be.

We have included chapters on seven specific eruptions, and we have described two areas—Hawaii and Iceland—where primordial volcanism reveals plate tectonics at work. In human terms these catastrophic events and titanic geological processes have led to consequences ranging from societal disruption to mass destruction, from ancient myths to modern movies. From the prodigious blast of Thera more than 3,600 years ago to the relative burp of Mount St. Helens in 1980, the volcanic events treated here, and their strings of aftereffects, attest to the many ways in which geological events and human destiny have been intertwined throughout history.

· Glossary

aa (*ah*-ah)—Hawaiian term for rough, clinkerlike lava (cf. pahoehoe).

aerosol (*air*-o-sol)—a suspension, in the air, of liquid droplets or of solid particles.

archipelago (ar-ke-*pel*-a-go)—a large group of islands, or an area of a sea that contains many islands.

ash (see volcanic ash).

asthenosphere (as-*theen*-o-sphere)—a ductile layer in the upper part of the earth's mantle (q.v.); it flows plastically and is where magma (q.v.) is thought to be generated (cf. lithosphere).

asthenosphere wedge (see mantle wedge).

atoll (*at*-all)—a coral reef that encircles, or nearly encircles, a central lagoon.

avalanche—a mass of rock, soil, ice, or snow, or a mixture of those materials, falling or sliding down a mountainside or hillside (cf. landslide).

basalt (ba-*salt*)—a dark-colored rock (solidified lava; q.v.) that contains little silicon.

base surge—a cloud of gas or suspended solid debris that rapidly moves outward from the base of an eruption column (cf. surge).

bomb (see volcanic bomb).

calcite (*kal*-site)—a common rock-forming mineral (q.v.) composed of calcium carbonate.

caldera (col-*dare*-uh)—a large, more or less circular volcanic depression, much larger than a volcanic crater (q.v.).

carbon-14—a radioactive carbon isotope (q.v.) that has a mass number of 14 instead of the normal 12.

chamber (see magma chamber).

cinder (volcanic)—a small fragment of material erupted by a volcano.

cinder cone—a conical hill formed by an accumulation of volcanic cinders and other debris.

conduit (see volcanic conduit).

cone (see cinder cone, spatter cone, volcanic cone).

coral—hard material composed of many calcium carbonate skeletons of tiny bottom-dwelling organisms (corals) in the sea.

core (of the earth)—the central part of the earth's interior, thought to be divided into a solid inner core and a fluid outer core.

core (see drill core, ice core).

crater (volcanic)—a depression, usually at the summit of a volcanic cone or mountain (cf. caldera).

crust (of the earth)—the thin, solid, outermost layer or shell of the earth.

dike (geological)—a tabular body of solidified magma that fills a fracture in surrounding rock.

dome (see volcanic dome).

drill core—a cylindrical portion of rock obtained by drilling with a hollow drill bit.

dry fog—a fog, or haze, composed of dry particulate matter rather than droplets of moisture.

earthquake focus—the point within the earth where rocks first rupture to cause an earthquake.

erosion—processes by which rocks and other materials at the earth's surface are worn away, as by abrasion, running water, solution, wind, or weathering (q.v.), the resulting materials being transported to a location different from where they originated.

eruption cloud—a cloud of gases and fragmental material created by a volcanic eruption, typically spreading out atop an eruption column (q.v.).

eruption column—a column of material forcefully ejected into the atmosphere during a volcanic eruption, before it spreads out to form an eruption cloud (q.v.).

explosivity—explosive power (see volcanic explosivity index).

fault (in rock)—a fracture or zone of fractures, one side of which has been displaced relative to the other in a direction parallel to the fracture (cf. thrust fault).

fault block—a unit of the earth's crust bounded by faults.

fault scarp—a steep slope or cliff formed by movement, usually up or down, along a fault.

fault zone—a zone, perhaps many hundreds of meters wide, containing numerous fractures.

fissure (in rock)—a fracture along which the two sides have separated (cf. radial fissure).

flank eruption—an eruption from the side, or flank, of a volcano, rather than from the summit.

focus (see earthquake focus).

fossil—any trace or remains of a plant or animal preserved in the earth's crust.

fracture (in rock)—any break, such as a fissure or fault (q.v.), caused by mechanical stress.

fumarole (*few*-ma-role)—a vent in the earth, usually volcanic, from which gases are emitted.

geochemical—pertaining to the distribution and amounts of chemical elements and their isotopes (q.v.) in the materials that make up the earth.

geology—the study of the planet earth and its history, including its life-forms, the materials of which it is made, the processes that act on those materials, and the results of those processes.

geophysicist—a scientist who studies the earth by using quantitative physical methods.

geyser—a hot spring that, from time to time, erupts jets of hot water and steam.

glacier—a large mass of ice formed by the compaction and recrystallization of snow, including both small mountain glaciers and continental ice sheets (q.v.).

glacier burst (see *jökulhlaup*).

gravity meter—an instrument for measuring variations in the earth's gravitation.

greenhouse effect—the heating of the earth's surface caused when outgoing heat is absorbed by water vapor and carbon dioxide in the atmosphere and subsequently returned to earth.

groundwater—water in the zone of saturation within the earth's crust.

guyot (goo-*yoh*)—a flat-topped seamount (q.v.).

hazard map—a map showing locations and types of hazards.

hot spot—a large volcanic center (q.v.) at the surface of the earth, thought to be caused by a plume of hot mantle (q.v.) material, which persists for many millions of years.

ice core—a cylinder of ice obtained by drilling with a hollow drill bit.

ice sheet—a glacier (q.v.) of considerable thickness that covers a large area of the earth's surface.

island arc (see volcanic arc).

isotope—one of two or more species of a chemical element that have different mass numbers (the sum of the number of protons and neutrons in the nucleus of an atom of the element).

jökulhlaup (*yo*-kul-laup)—Icelandic term for a sudden release of meltwater (q.v.) from a glacier; also called a glacier burst.

jökull (*yo*-kul)—Icelandic term for a glacier.

landslide—downslope movement of large amounts of soil and rock material.

lapilli (la-*pill*-ee)—small fragments of material ejected from a volcano.

lava—rock erupted from a volcano or volcanic fissure, either in its original molten state or after it has cooled and hardened (cf. magma).

lava flow—an outpouring of lava from a volcanic vent or fissure, either molten or solidified (see volcano).

lava tube—a longitudinal hollow space beneath the surface of a solidified lava flow, formed when molten lava continues to flow after the surface lava has cooled to form a crust.

limestone—a type of rock made up mostly of calcium carbonate.

lithosphere (*lith*-o-sphere)—a layer of the earth comprising the crust and the uppermost, solid part of the mantle (q.v.); it resists plastic flow and lies directly above the asthenosphere (q.v.).

magma (*mag*-ma)—molten rock generated within the earth (cf. lava).

magma chamber—a reservoir of magma within the lithosphere (q.v.).

magmatic (mag-*mat*-ic)—of, pertaining to, or derived from magma.

magnitude (of earthquakes)—a measure of the strain energy released by an earthquake (see Richter scale).

mantle (of the earth)—the part of the earth between the crust and the core, comprising most of the earth's volume.

mantle plume—a localized body of molten rock rising from the earth's mantle into the crust, presumably the cause of a hot spot (q.v.).

mantle wedge—wedge of mantle material that occupies the space between one tectonic plate and another that is sliding beneath it (see plate tectonics).

marble—a metamorphic (q.v.) rock consisting mostly of recrystal-lized calcite (q.v.); specifically, metamorphic limestone (q.v.).

massif (ma-*seef*)—a massive topographic feature, such as a large mountain or group of mountains forming part of a mountain range.

meltwater—water derived from the melting of snow or ice; espe-cially a stream flowing in, under, or from a melting glacier (q.v.).

metamorphic (met-a-*mor*-fic)—pertaining to or produced by the alteration of rock by physical and chemical conditions at depth within the earth's crust.

meteorologist—a scientist who studies the earth's atmosphere.

mineral—a naturally occurring inorganic element or compound with characteristic chemical composition, crystal form, and physical properties.

mudflow—a flowing mass of mostly fine-grained earth material mixed with water.

nuée ardente (*new*-ay ar-*dahnt*) (see pyroclastic flow).

oceanic ridge—in mid-ocean, a high, broad swelling of the sea floor, with a central rift valley (q.v.); the site of sea-floor spreading (q.v.).

oceanic rift—a rift (q.v.) in the ocean floor.

oceanic trench—a deep, narrow, longitudinal depression between the margin of a continental shelf and the floor of the deep sea beyond.

outwash plain—a broad, gently sloping area of material deposited by meltwater (q.v.) from a glacier.

pahoehoe (pa-*hoy-hoy*)—Hawaiian term for lava that has a smooth surface, sometimes undulating and sometimes resembling twisted strands of rope.

paleomagnetism—the strength and direction of the earth's mag-netic field during the geologic past.

pali (*pal*-ee)—Hawaiian term for a steep slope or escarpment.

Pele (*pay*-lay)—ancient Hawaiian goddess of fire.

Peléan (pay-*lay*-an) eruption—a volcanic eruption that produces pyroclastic flows (q.v.) and volcanic domes (q.v.); named for Mount Pelée on the Caribbean island of Martinique (not connected with the Hawaiian fire goddess Pele).

Pele's hair—naturally occurring thin strands of volcanic glass (q.v.) formed in fountains of fluid lava; named for the Hawaiian fire goddess Pele.

Pele's tears—solidified droplets of volcanic glass (q.v.); named for the Hawaiian fire goddess Pele.

permeability—the capacity of porous rock, sediment, or soil to transmit a fluid, such as water.

plastic—capable of being deformed without rupturing.

plate (see plate tectonics).

platelet (*plate*-let)—a small tectonic plate (see plate tectonics).

plate tectonics—the theory that segments of the earth's lithosphere (q.v.), called tectonic plates, move about, giving rise to earthquakes and volcanic activity.

Plinian (*plin*-ee-an) **eruption**—an explosive volcanic eruption in which a stream of fragmental materials and gases bursts from a vent at high velocity, characteristically with a high eruption column (q.v.); named for Pliny the Younger, who in 79 c.e. first described such an eruption, from Mount Vesuvius.

plume (see mantle plume).

pumice (*pum*-iss)—a light-colored volcanic rock made frothy and buoyant by gas bubbles.

pyroclastic (py-ro-*clas*-tic)—pertaining to fragmental, or clastic, rock material created during a volcanic explosion.

pyroclastic flow—a hot, swiftly moving, turbulent gaseous cloud, sometimes incandescent, that erupts from a volcano and contains fragments of volcanic rock.

radial fissure—one of a group of fissures that radiate from a central point, as around a volcanic vent.

Richter scale—a logarithmic scale of earthquake magnitude, devised in 1935 by seismologist Charles Francis Richter.

rift (in the earth's crust)—a long, narrow regional trough, bounded by fault zones (q.v.), where the lithosphere (q.v.) has ruptured by being pulled apart.

rift valley—the valley within a rift, specifically the deep central cleft along the crest of an oceanic ridge (q.v.).

Ring of Fire—informal name given to the belts of volcanic and seismic activity that virtually surround the Pacific Ocean.

scarp (see fault scarp).

scoria—highly irregular, cinderlike volcanic material, similar to pumice (q.v.) but heavier and darker.

sea-floor spreading—movement of the sea floor away from either side of an oceanic rift (q.v.) as magma (q.v.), welling up through the rift, continually creates new oceanic crust; sea-floor spreading is related to the dynamic movement of tectonic plates (see plate tectonics).

seamount—a mountain that rises from the sea floor but does not reach the surface of the water (cf. guyot).

seismic—pertaining to earthquakes and earth vibrations, whether natural or artificial in origin.

seismograph—an instrument for detecting and recording vibrations in the earth, especially earthquakes.

seismologist—a scientist who studies earthquakes and the earth's internal structure.

shield volcano—a volcano shaped like a flattened dome (or warrior's shield), created by solidified lava flows of low viscosity (thus high fluidity), as in the Hawaiian Islands.

sill (geological)—a tabular body of solid rock that, while molten, flowed between layers of surrounding rock.

slag—glassy waste material produced by the smelting of metallic ore.

spatter cone—a low, steep-sided cone formed by the hardening of molten fragmental materials around a volcanic vent.

spreading (see sea-floor spreading).

strain—deformation of a solid material as a result of stress.

stress—force exerted per unit area of a solid material.

subaerial—pertaining to events or processes that occur on or near a land surface (cf. submarine).

subduction—the descent of one tectonic plate beneath another (see plate tectonics).

submarine (adjective)—pertaining to events or processes that occur beneath the surface of a body of water (cf. subaerial).

sulfur dioxide—a gaseous chemical compound comprising one part sulfur to two parts oxygen.

surge (volcanic)—a turbulent cloud of fine volcanic debris and fiery gases.

tectonic (tek-*tahn*-ic)—pertaining to structural and deformational features within the outer part of the earth (see plate tectonics).

tectonic plate (see plate tectonics).

thrust fault—a fault (q.v.) caused by the thrusting of one part of the earth's crust over or under another.

tiltmeter—an instrument for measuring slight changes in the slope, or tilt, of the earth's surface.

topography—the three-dimensional configuration of a land surface.

trench (see oceanic trench).

tsunami (tsu-*nom*-ee)—a sea wave produced by a sudden disturbance of the ocean floor, as by an earthquake, or by volcanic activity, as when a pyroclastic flow (q.v.) slams into the sea.

VEI (see volcanic explosivity index).

viscous—having relatively high viscosity, or resistance to flow.

volcanic arc—a generally curved linear belt of volcanoes or volcanic islands, such as those that form above a zone of subduction (q.v.).

volcanic ash—fine fragments of material ejected from a volcano.

volcanic belt (see volcanic arc).

volcanic bomb—a body of viscous material that is ejected from a volcano while molten and thus attains a rounded shape while in flight.

volcanic center—a site at which volcanic activity is occurring or has occurred in the past.

volcanic complex—a group of volcanic centers.

volcanic conduit—the channel through which magma (q.v.) rises into a volcano from within the earth's crust.

volcanic cone—a relatively small, conical hill of lava (q.v.) or fragmental material that has been built up around a vent in the earth's surface (cf. cinder cone).

volcanic dome—a steep-sided mass of highly viscous lava (q.v.) extruded from a volcano to form a dome-shaped structure above and around the volcano's vent.

volcanic explosivity index (VEI)—a measure, similar in principle to the Richter scale (q.v.) of earthquake intensities, by which ten classes of volcanic activity are distinguished, each higher class representing a tenfold increase in explosivity, or explosive power, ranging from gently effusive lava flows (q.v.) to highly explosive, cataclysmic eruptions.

volcanic gas—a natural gas, comprising mostly water vapor (steam) and also including carbon dioxide, sulfur dioxide, hydrogen sulfide, and smaller amounts of other gases, dissolved in magma (q.v.) and released during a volcanic eruption.

volcanic glass—a natural glass produced when molten lava cools too rapidly for the material to form crystals.

volcanic vent (see volcano).

volcanism—the process by which magma (q.v.) and associated gases rise through the earth's crust and are extruded onto the surface or into the atmosphere.

volcano—a vent in the earth's surface through which magma (q.v.), associated gases, and fragmental material can erupt; technically includes fissures (q.v.) as well as volcanic cones (q.v.) and volcanic mountains, but here, for simplicity, we limit the term to volcanic mountains.

volcanology—the branch of geology that deals with volcanoes and volcanism.

weathering—the process by which materials exposed to the atmosphere are changed in color, texture, composition, firmness, or form; specifically the disintegration and chemical decomposition of rock (cf. erosion).

· Notes and References

PREFACE

Notes

1. Snow, C. P., *The Two Cultures* (Cambridge: Cambridge University Press, 1959).
2. Trevanian, *The Summer of Katya* (New York: Crown Publishers, 1983), 99.

1 · VOLCANISM: ORIGINS AND CONSEQUENCES

Notes

1. Krafft and Krafft, *Volcano*, 83.
2. Ibid., 59.
3. Grosvenor, "Hawaiian Islands," 235.
4. Endo, *Volcano*, 27.
5. Simkin and Siebert, *Volcanoes*, 10.

Cited References

Endo, Shusaku. *Volcano.* Translated by Richard A. Schuchert. New York: Taplinger, 1980.

Grosvenor, Gilbert. "The Hawaiian Islands." *National Geographic* 45, no. 2 (1924): 115–238.

Krafft, Maurice, and Katia Krafft. *Volcano.* New York: Harry N. Abrams, 1975.

Simkin, Tom, and Lee Siebert. *Volcanoes of the World: A Regional Directory, Gazetteer, and Chronology of Volcanism during the Last 10,000 Years.* 2d ed. Tucson, Ariz.: Geoscience Press, 1994.

Related Reading

Decker, Robert, and Barbara Decker. *Volcanoes.* 3d ed. New York: W. H. Freeman, 1997.

Fisher, Richard V., Grant Heiken, and Jeffrey B. Hulen. *Volcanoes: Crucibles of Change*. Princeton, N.J.: Princeton University Press, 1997.

Francis, Peter. *Volcanoes: A Planetary Perspective*. Oxford: Clarendon Press and Oxford University Press, 1993.

Knight, Linsay. *Volcanoes and Earthquakes*. Edited by Eldridge M. Moores. Alexandria, Va.: Time Life Books, 1996.

McClelland, Lindsay, Tom Simkin, Marjorie Summers, Elizabeth Nielsen, and Thomas C. Stein, eds. *Global Volcanism, 1975–1985: The First Decade of Reports from the Smithsonian Institution's Scientific Event Alert Network (SEAN)*. Englewood Cliffs, N.J.: Prentice-Hall; Washington, D.C.: American Geophysical Union, 1989.

Sigurdsson, Haraldur, Bruce F. Houghton, Stephen R. McNutt, Hazel Rymer, and John Stix, eds. *Encyclopedia of Volcanoes*. New York: Academic Press, 2000.

Wood, Charles A., and Jürgen Kienle. *Volcanoes of North America: United States and Canada*. Cambridge: Cambridge University Press, 1990.

2 • THE HAWAIIAN ISLANDS AND THE LEGACY OF PELE THE FIRE GODDESS

Notes

1. Twain, *Roughing It*, 263.

2. Twain, *Letters from Hawaii*, vi.

3. McPhee, "Control of Nature," 70.

4. Darwin, *On the Structure and Distribution of Coral Reefs*.

5. Twain, *Letters from Hawaii*, 295–97.

6. Westervelt, *Hawaiian Legends of Volcanoes*, ix–x.

7. Ibid., 180.

8. Day, "Liholiho and the Longnecks," 146.

9. Tennyson, *Works of Alfred Lord Tennyson*, 5:65–68.

Cited References

Darwin, Charles R. *On the Structure and Distribution of Coral Reefs; Also Geological Observations on the Volcanic Islands and Parts of South America Visited during the Voyage of H.M.S. Beagle*. London: Ward, Lock, 1890.

Day, A. Grove. "Liholiho and the Longnecks." In *A Hawai'i Anthology*, ed. Joseph Stanton, 142–47. Honolulu: State Foundation on Culture and the Arts, 1997.

McPhee, John. "The Control of Nature." *The New Yorker*, February 29, 1988, 70.

Tennyson, Alfred, Lord. *The Works of Alfred Lord Tennyson.* Vol. 5. Edited by Hallam, Lord Tennyson. New York: Macmillan, 1908.

Twain, Mark. *Mark Twain's Letters from Hawaii.* Edited by A. Grove Day. New York: Appleton-Century, 1966.

———. *Roughing It.* New York: Harper and Brothers, 1871.

Westervelt, William D. *Hawaiian Legends of Volcanoes.* Boston: Charles E. Tuttle, 1963.

Other Sources

Brantley, Steven R. *Volcanoes of the United States.* General Interest Publication, U.S. Geological Survey, 1999.

Dalrymple, G. B., M. A. Lanphere, and D. A. Clague, *Ages of Emperor Seamounts Confirm Hot-Spot Hypothesis.* Professional Paper, U.S. Geological Survey, 1979.

Dibble, Sheldon. *History and General Views of the Sandwich Islands' Mission.* New York: Taylor & Dodd, 1839.

Kamakau, Samuel M. *Ka Po'e Kahiko: The People of Old.* Translated by Mary Kawena Pukui and edited by Dorothy B. Barrère. Bernice P. Bishop Museum special publication, no. 51. Honolulu: Bishop Museum Press, 1968.

Kauahikoua, Jim, Tom Hildebrand, and Mike Webring. "Deep Magmatic Structures of Hawaiian Volcanoes; Imaged by Three-Dimensional Gravity Models." *Geology* 28, no. 10 (2000): 883–86.

Keating, Barbara. *Hawaiian Eruptions: The Eruptions of Kilauea and Mauna Loa Volcanoes.* Honolulu: University of Hawaii Foundation, 1987.

Lassiter, John. "Hawaiian Plume Dynamics." *Science* 285, no. 5429 (1999): 841–47.

Moore, James G., William R. Normark, and Robin T. Holcomb. "Giant Hawaiian Underwater Landslides." *Science* 264, no. 5155 (1994): 46–47.

Stearns, Harold T. *Geology of the State of Hawaii.* 2d ed. Palo Alto, Calif.: Pacific Books, 1985

Related Reading

MacDonald, Gordon A., and Agatin T. Abbott. *Volcanoes in the Sea: The Geology of Hawaii.* Honolulu: University of Hawaii, 1970.

3 • THE BRONZE AGE ERUPTION OF THERA: DESTROYER OF ATLANTIS AND MINOAN CRETE?

Notes

1. Manning, "Thera Eruption," 97.

2. Minoura et al., "Discovery of Minoan Tsunami Deposits."

3. Durant, *Story of Civilization,* 2:5.

4. Hamilton and Cairns, *Collected Dialogues of Plato,* 1159–60.

5. Ibid., 1218.

6. Ibid., 1220.

7. Galanopoulos and Bacon, *Atlantis,* 133–34.

Cited References

Durant, Will. *The Story of Civilization.* Vol. 2, *The Life of Greece.* New York: Simon and Schuster, 1939.

Galanopoulos, Angelos Georgiou, and Edward Bacon. *Atlantis: The Truth behind the Legend.* Indianapolis, Ind.: Bobbs-Merrill, 1969.

Hamilton, Edith, and Huntington Cairns, eds. *The Collected Dialogues of Plato, Including the Letters.* Translated by Lane Cooper et al. Bollingen series, no. 71. Princeton, N.J.: Princeton University Press, 1989.

Hull, Edward. *Volcanoes: Past and Present.* London: Walter Scott, Ltd., 1892.

Manning, S. W. "The Thera Eruption: The Third Congress and the Problem of the Date." *Archaeometry* 32, no. 1 (1990): 1–100.

Minoura, K., F. Imamura, U. Kuran, T. Nakamura, G. A. Papadopoulos, T. Takahashi, and C. A. Yalciner. "Discovery of Minoan Tsunami Deposits." *Geology* 28, no. 1 (2000): 59–62.

Other Sources

Doumas, Christos G. *Thera: Pompeii of the Ancient Aegean.* London: Thames and Hudson, 1983.

Durant, Will. *The Story of Civilization.* Vol. 1, *Our Oriental Heritage.* New York: Simon and Schuster, 1954.

Fouqué, Ferdinand A. *Santorini and Its Eruptions.* Translated by Alexander R. McBirney. Baltimore: Johns Hopkins University Press, 1998.

Friedrich, Walter L. *Fire in the Sea: The Santorini Volcano: National History and the Legend of Atlantis.* Translated by Alexander R. McBirney. Cambridge: Cambridge University Press, 2000.

Hardy, D. A., and A. C. Renfrew, eds. *Thera and the Aegean World III: Proceedings of the Third International Congress, Santorini, Greece, 3–9 September 1989.* 3 vols. London: Thera Foundation, 1990.

La Moreaux, P. E. "Worldwide Environmental Impacts of the Eruption of Thera." *Environmental Geology* 26, no. 3 (1995): 172–81.

Luce, J. V. *Lost Atlantis: New Light on an Old Legend.* New York: McGraw-Hill, 1969.

McCoy-Floyd, W., and G. Heiken. "Tsunami Generated by the Late Bonze Age Eruption of Thera (Santorini)." *Pure and Applied Geophysics* 157, no. 6–8 (2000): 1227–56.

Page, D. L. *The Santorini Volcano and the Destruction of Minoan Crete.* London: Society for the Promotion of Hellenic Studies, 1970.

Pellegrino, Charles. *Unearthing Atlantis.* New York: Vintage Books (Random House), 1993.

Pyle, D. M. "The Application of Tree Ring and Ice Core Studies to the Dating of the Minoan Eruption." In *Thera and the Aegean World III: Proceedings of the Third International Congress, Santorini, Greece, 3–9 September 1989,* ed. D. A. Hardy and A. C. Renfrew, 2:167–73. London: Thera Foundation, 1990.

———. "Ice Core Acidity Peaks, Retarded Tree Growth, Putative Eruptions." *Archaeometry* 31 (1989): 88–91.

———. "New Estimates for the Volume of the Minoan Eruption." In *Thera and the Aegean World III: Proceedings of the Third International Congress, Santorini, Greece, 3–9 September 1989,* ed. D. A. Hardy and A. C. Renfrew, 2:113–21. London: Thera Foundation, 1990.

Vitaliano, Dorothy B. *Legends of the Earth.* Secaucus, N.J.: Citadel Press, 1976.

Related Reading

Druitt, T. H., L. Edwards, R. M. Mellors, D. M. Pyle, R.S.J. Sparks, M. Lanphere, M. Davies, and B. Barriero. *Santorini Volcano.* Geological Society Memoir, no. 19. London: Geological Society Publishing House, 2000.

4 • THE ERUPTION OF VESUVIUS IN 79 C.E.: CULTURAL REVERBERATIONS THROUGH THE AGES

Notes

1. Stothers and Rampino, "Volcanic Eruptions in the Mediterranean," 6360.

2. Strabo, *Geography of Strabo,* 2:453

3. Dio Cassius, *Dio's Roman History,* 8:305–7.

4. Pliny the Younger, *Letters of the Younger Pliny,* 166. Interestingly, other translations of Pliny do not use the term *umbrella pine,* which is so descriptive of a common species of Italian pine trees. Instead, they simply use the term *pine tree*—but the tree in question surely is *Pinus pinea,* the Mediterranean stone pine, with its tall, often branchless trunk and its rounded canopy of foliage at the top.

5. Ibid., 171.

6. Hoffer, *Volcano,* 130.

7. Morris, *Volcano's Deadly Work,* 258–59.

8. Ibid., 260.

9. Malaparte, *The Skin*, 257–59.

10. Lockhart, *Memoirs of the Life of Sir Walter Scott*, 5:397; Dickens, *Pictures from Italy*, 224–25, 227.

11. Goethe, *Italian Journey*, 206–7.

12. Stendhal, *Rome, Naples, and Florence*, 371.

13. Twain, *Traveling with the Innocents Abroad*, 76–77.

14. James, *Italian Hours*, 358.

15. Lobley, *Mount Vesuvius*, 16.

16. Shelley, *Poetical Works of Percy Bysshe Shelley*, 42.

17. Sontag, *Volcano Lover*, 112.

Cited References

Alfano, G. B., and L. Friedlaender. *Die Geschichte des Vesuv*. Berlin: Verlag Dietrich Reimer (E. Vohsen), A.-G., 1929.

Bullard, Fred M. *Volcanoes of the Earth*. Rev. ed. Austin: University of Texas Press, 1962.

Dickens, Charles. *Pictures from Italy*. New York: Coward, McCann & Geoghegan, 1974.

Dio Cassius. *Dio's Roman History*. Translated by Earnest Cary. Vol. 8. The Loeb Classical Library, no. 176. London: William Heinemann, 1925.

Goethe, Johann Wolfgang. *Italian Journey*. Translated by W. H. Auden and Elizabeth Mayer. New York: Pantheon Books, 1962.

Hoffer, William. *Volcano: The Search for Vesuvius*. New York: Summit Books, 1982.

James, Henry. *Italian Hours*. Westport, Conn.: Greenwood Press, 1977.

Lobley, J. Logan. *Mount Vesuvius*. London: Roper and Drowley, 1889.

Lockhart, John Gibson. *Memoirs of the Life of Sir Walter Scott*. Vol. 5. New York: Houghton Mifflin, 1902.

Malaparte, Curzio. *The Skin*. Boston: Houghton Mifflin, 1952.

Morris, Charles. *The Volcano's Deadly Work: From the Fall of Pompeii to the Destruction of St. Pierre*. W. E. Scull, 1902.

Pliny the Younger. *The Letters of the Younger Pliny*. Translated by Betty Radice. Harmondsworth, England: Penguin Books, 1963.

Shelley, Percy Bysshe. *The Poetical Works of Percy Bysshe Shelley*. London: Reeves and Turner, 1882.

Sigurdsson, Haraldur, S. Carey, W. Cornell, and T. Pescatore. "The Eruption of Vesuvius in 79 A.D." *National Geographic Research* 1 (1985): 332–87.

Sontag, Susan. *The Volcano Lover: A Romance.* New York: Farrar Straus Giroux, 1992.

Stendhal. *Rome, Naples, and Florence.* Translated by Richard N. Coe. London: John Calder, 1959.

Stothers, Richard B., and Michael R. Rampino. "Volcanic Eruptions in the Mediterranean before A.D. 630." *Journal of Geophysical Research* 88, no. B8 (1983): 6357–71.

Strabo. *The Geography of Strabo.* Vol. 2. Translated by Horace Leonard Jones. The Loeb Classical Library, no. 50. London: William Heinemann, 1923.

Twain, Mark. *Traveling with the Innocents Abroad.* Edited by Daniel Morley McKeithan. Norman: University of Oklahoma Press, 1958.

Other Sources

Bowerstock, G. W. "On the 24th of August, A.D. 79, Antiquity Laid Down a Unique Legacy for the Future." *Harvard Magazine,* May–June 1978, 40–49.

Brion, Marcel. *Pompeii and Herculaneum: The Glory and the Grief.* Translated by John Rosenberg. London: Elek Books, 1960.

Dayton, Leigh. "The Fat, Hairy Women of Pompeii." *New Scientist* 143 (September 24, 1994): 10.

Durant, Will. *The Story of Civilization.* Vol. 3, *Caesar and Christ.* New York: Simon and Schuster, 1944.

Gasparini, P. "Looking inside Mount Vesuvius." *Eos* 79, no. 19 (1998): 229–30.

Gore, Rick. "A Prayer for Pozzuoli." *National Geographic* 165, no. 5 (1984): 614–25.

———. "The Dead Do Tell Tales at Vesuvius." *National Geographic* 165, no. 5 (1984): 557–613.

Gould, Stephen Jay. "Lyell's Pillars of Wisdom." *Natural History* 108 (April 1999): 28–34, 87–89.

———. "Pozzuoli's Pillars Revisited." *Natural History* 108 (May 1999): 24, 81–83, 88, 90–91.

Hartt, Frederick. *Art: A History of Painting, Sculpture, Architecture.* 4th ed. Englewood Cliffs, N.J.: Prentice-Hall; New York: Harry N. Abrams, 1993.

Heiken, Grant. "Will Vesuvius Erupt? Three Million People Need to Know." *Science* 286, no. 5445 (1999): 1685–87.

Judge, Joseph. "A Buried Roman Town Gives Up Its Dead." *National Geographic* 162, no. 6 (1982): 687–93.

Kruger, Christoph, ed. *Volcanoes.* New York: G. P. Putnam's Sons, 1971.

Maiuri, Amedeo. "Last Moments of the Pompeiians." *National Geographic* 120, no. 5 (1961): 651–69.

Picard, M. "Vesuvius." *Journal of Geoscience Education* 48, no. 4 (2000): 533–40.

Pompeii as Source and Inspiration: Reflections in Eighteenth- and Nineteenth-Century Art. Report of an exhibition organized by the 1976–1977 graduate students in the Museum Practice Program. Ann Arbor: Museum of Art, University of Michigan, 1977.

Rosi, Mauro, Claudia Pincipe, and Raffaella Vecci. "The 1631 Vesuvius Eruption. A Reconstruction Based on Historical and Stratigraphical Data." *Journal of Volcanology and Geothermal Research* 58 (1993): 151–282.

Scandone, Roberto, Lisetta Giacomelli, and Paolo Gasparini. "Mount Vesuvius: 2000 Years of Volcanological Observations." *Journal of Volcanology and Geothermal Research* 58 (1993): 5–25.

Related Reading

Bulwer-Lytton, Sir Edward. *The Last Days of Pompeii.* Garden City, N.Y.: International Collectors Library, 1946.

Feder, Theodore H. *Great Treasures of Pompeii and Herculaneum.* New York: Abbeville Press, 1978.

Lloyd, Alan. *Alive in the Last Days of Pompeii.* Los Angeles: Pinnacle Books, 1975.

Vanags, Patricia. *The Glory That Was Pompeii.* New York: Mayflower Books.

For Viewing

National Geographic Society. *In the Shadow of Vesuvius.* 1989. Videocassette.

5 · ICELAND: COMING APART AT THE SEAMS

Notes

1. Pfeiffer, *Visit to Iceland*, 337.

2. Colum, *Orpheus*, 208.

3. Carlyle, *On Heroes, Hero-Worship, and the Heroic in History*, 23.

4. Thorarinsson, *Eruption of Hekla*, 26–27.

5. Thorarinsson, *Hekla*, 6.

6. Bárdarson, *Ice and Fire*, 16.

7. Scarth, *Vulcan's Fury*, 112.

8. Franklin, "Meteorological Imaginations and Conjectures," 8:488–89.

9. Wood, "Amazing and Portentous Summer of 1783," 410.

10. Thorarinsson, *Surtsey*, 9–10.

11. Ibid., 19.

12. Hayes, *Island on Fire*, 147.

13. Gunnarsson, *Volcano*, 29.

14. Colgate and Sigurgeirsson, "Dynamic Mixing of Water and Lava," 552.

Cited References

Bárdarson, Hjálmar R. *Ice and Fire: Contrast of Icelandic Nature.* Reykjavik: Hjálmar Bárdarson, 1971.

Carlyle, Thomas. *On Heroes, Hero-Worship, and the Heroic in History.* New York: D. Appleton, 1842.

Colgate, S. A., and Thorbjörn Sigurgeirsson. "Dynamic Mixing of Water and Lava." *Nature* 244 (August 31, 1973): 552–55.

Colum, Padraic. *Orpheus: Myths of the World.* New York: Macmillan, 1930.

Franklin, Benjamin. "Meteorological Imaginations and Conjectures." In *The Complete Works of Benjamin Franklin,* comp. and ed. John Bigelow, 8:486–89. New York: G. P. Putnam's Sons, 1888.

Gunnarsson, Árni. *Volcano: Ordeal by Fire in Iceland's Westmann Islands.* Translated by May and Hallberg Hallmundsson. Reykjavik: Iceland Review, 1973.

Hayes, Joseph. *Island on Fire.* New York: Grosset & Dunlap, 1979.

Lawver, Lawrence A., and R. D. Mueller. "Iceland Hot Spot Track." *Geology* 22, no. 4 (1994): 311–14.

Pfeiffer, Ida. *A Visit to Iceland.* London: Ingram, Cooke, 1852.

Scarth, Alwyn. *Vulcan's Fury: Man against the Volcano.* New Haven, Conn.: Yale University Press, 1999.

Thorarinsson, Sigurdur. *The Eruption of Hekla, 1947–1948.* Reykjavik: H. F. Leiftur, 1967.

———. *Hekla: A Notorious Volcano.* Reykjavik: Almenna Bókafélagid, 1970.

———. *Surtsey.* New York: Viking Press, 1967.

Wood, Charles A. "Amazing and Portentous Summer of 1783." *Eos* 65, no. 26 (1984): 410–11.

Other Sources

Bárdarson, Hjálmar R. *Iceland: A Portrait of Its Land and People.* Reykjavik: H. R. Bárdarson, 1989.

Brayshay, Mark, and John Grattan. "Environmental and Social Responses in Europe to the 1783 Eruption of the Laki Fissure: A Consideration of Contemporary Documentary Evidence." In *Volcanoes in the Quaternary,* ed. C. R. Firth and W. J. McGuire, 173–87. Geological Society special publication, no. 161. London: Geological Society, 1999.

Clague, J. L., and W. H. Mathews. "The Magnitude of Jökulhlaups." *Journal of Glaciology* 12 (1973): 501–4.

Jacoby, Gordon C., Karen W. Workman, and Roseanne D. Arrigo. "Laki Eruption of 1783, Tree Rings and Disaster for Northwest Alaska Inuit." *Quaternary Science Reviews* 18, no. 12 (1999): 1365–71.

Kingston, J. A. *The Weather Patterns for the 1780s over Europe.* Cambridge: Cambridge University Press, 1988.

McPhee, John. "The Control of Nature: Cooling the Lava." Part 1. *The New Yorker,* February 22, 1988, 43–77.

———. "The Control of Nature: Cooling the Lava." Part 2. *The New Yorker,* February 29, 1988, 64–79.

Pyle, David M. "How Did the Summer Go?" *Nature* 393, no. 6684 (1998): 415–17.

Sigurdsson, Haraldur. "Volcanic Pollution and Climate: The 1783 Laki Eruption." *Eos* 63 (1982): 601–2.

Stothers, Richard B. "Volcanic Dry Fogs, Climate Cooling, and Plague Pandemics in Europe and the Middle East." *Climatic Change* 42, no. 4 (1999): 713–23.

Thorarinsson, Sigurdur. "The Lakagigar Eruption of 1783." *Bulletin Volcanologique* 33 (1969): 910–27.

Thordarson, Th., and S. Self. "The Laki (Skaftár Fires) and Grimsvötn Eruptions." *Bulletin of Volcanology* 55 (1993): 233–63.

Wood, Charles A. "Climatic Effects of the 1783 Laki Eruption." In *The Year without a Summer? World Climate in 1816,* ed. C. R. Harington, 58–73. Ottawa: Canadian Museum of Nature, 1992.

6 · THE ERUPTION OF TAMBORA IN 1815 AND "THE YEAR WITHOUT A SUMMER"

Notes

1. Lyell, *Principles of Geology,* 2:104–5.

2. Zollinger, *Besteigung des Vulkanes Tambora,* 5.

3. Arnold, *Famine,* 30.

4. Byron, *Complete Poetical Works,* 189.

5. Shelley, *Frankenstein,* 20–23.

6. Stommel and Stommel, *Volcano Weather,* 37.

7. Pike, *Granite Laughter and Marble Tears,* 32.

8. Stommel and Stommel, *Volcano Weather,* 108.

Cited References

Arnold, David J. *Famine: Social Crisis and Historical Change.* Oxford: Oxford University Press, 1988.

Byron, George Gordon, Baron. *The Complete Poetical Works of Byron.* Boston: Houghton Mifflin, 1905.

Lyell, Charles. *Principles of Geology.* 10th ed. Vol. 2. London: John Murray, 1868.

Pike, Robert E. *Granite Laughter and Marble Tears.* Stephen Daye Press, 1938.

Self, S., M. R. Rampino, M. S. Newton, and J. A. Wolf. "Volcanological Study of the Great Tambora Eruption of 1815." *Geology* 12, no. 12 (1984): 659–63.

Shelley, Mary Wollstonecraft. *Frankenstein, or The Modern Prometheus.* London: Oxford University Press, 1969.

Stommel, Henry, and Elizabeth Stommel. *Volcano Weather.* Newport, R.I.: Seven Seas Press, 1983.

Zollinger, Heinrich. *Besteigung des Vulkanes Tambora auf der Insel Sumbawa und Schilderung der Erupzion desselben in Jahr 1815.* Winterthur, Switzerland: Joh. Wurster & Comp., 1855.

Other Sources

Delmas, R. J., S. Kirchner, J. M. Palais, P. J. Robert. "1000 Years of Explosive Volcanism Recorded at the South Pole." *Tellus,* ser. B, *Chemical and Physical Meteorology* 44, no. 4 (1992): 335–50.

Harington, C. R., ed. *The Year without a Summer: World Climate in 1816.* Ottawa: Canadian Museum of Nature, 1992.

Pyle, David M. "How Did the Summer Go?" *Nature* 393, no. 6684 (1998): 415–17.

Sigurdsson, Haraldur, and Steven Carey. "Eruptive History of Tambora Volcano, Indonesia." In *The Sea off Mount Tambora,* ed. E. T. Degens, H. K. Wong, and M. T. Zen, 187–206. Mitteilungen aus dem Geologisch-Paläontologischen Institut der Universität Hamburg, no. 70. Hamburg: Im Selbstverlag des Geologisch-Paläontologischen Instituts der Universität Hamburg, 1992.

———. "Generation and Dispersal of Tephra from the 1815 Eruption of Tambora Volcano, Indonesia." In *The Sea off Mount Tambora,* ed. E. T. Degens, H. K. Wong, and M. T. Zen, 207–26. Mitteilungen aus dem Geologisch-Paläontologischen Institut der Universität Hamburg, no. 70. Hamburg: Im Selbstverlag des Geologisch-Paläontologischen Instituts der Universität Hamburg, 1992.

———. "Plinian and Co-ignimbrite Tephra Fall from the 1815 Eruption of Tambora Volcano." *Bulletin Volcanologique* 51 (1989): 243–70.

Stothers, Richard B. "The Great Tambora Eruption and Its Aftermath." *Science* 224, no. 4654 (1984): 1191–98.

———. "Volcanic Dry Fogs, Climate Cooling, and Plague Pandemics in Europe and the Middle East." *Climatic Change* 42, no. 4 (1999): 713–23.

7 · Krakatau, 1883:
Devastation, Death, and Ecologic Revival

Notes

1. Judd, "Earlier Eruptions of Krakatoa," 365.
2. Simkin and Fiske, *Krakatau 1883*, 73.
3. Ibid., 44.
4. Verbeek, *Krakatau*, pt. 1, p. iv.
5. Simkin and Fiske, *Krakatau 1883*, 117.
6. Ricks, *Poems of Tennyson*, 225.
7. Krafft and Krafft, *Volcano*, 140.

Cited References

Judd, John W. "The Earlier Eruptions of Krakatoa" [Letter to the editor]. *Nature* 40 (August 15, 1889): 365–66.

Krafft, Maurice, and Katia Krafft. *Volcano*. New York: Harry N. Abrams, 1975.

The Motion Picture Guide. Chicago: Cinebooks, Inc., 1985.

Ricks, Christopher, ed. *The Poems of Tennyson*. London: Longmans, Green, 1969.

Simkin, Tom, and Richard S. Fiske. *Krakatau 1883: The Volcanic Eruption and Its Effects*. Washington, D.C.: Smithsonian Institution Press, 1983.

Strachey, R. "The Krakatau Airwave." *Nature* 29 (1883): 181–83.

Verbeek, R.D.M. *Krakatau* (in French). Batavia: Imprimerie de l'état, 1886.

Yokoyama, I. "A Scenario of the 1883 Krakatau Tsunami," *Journal of Volcanology and Geothermal Research* 34, no. 1–2 (1987): 123–32.

Other Sources

Boutelle, C. O. "Water Waves from Krakatoa" [Letter to the editor]. *Science* 3, no. 73 (1884): 777.

Carey, Steven, David Morelli, Haraldur Sigurdsson, and Bronto Sutikno. "Tsunami Deposits from Major Explosive Eruptions: An Example from the 1883 Eruption of Krakatau." *Geology* 29, no. 4 (2001): 347–50.

Divers, Edward. "The Remarkable Sunsets" [Letter to the editor]. *Nature* 29 (January 24, 1884): 283–84.

Forbes, H. O. "The Volcanic Eruption of Krakatoa." *Proceedings of the Royal Geographical Society* 7 (1884): 142–52.

Hopkins, Gerard. "The Remarkable Sunsets" [Letter to the editor]. *Nature* 29 (January 3, 1884): 222–23.

Judd, John W. "The Dust of Krakatoa" [Letter to the editor]. *Nature* 29 (April 24, 1984): 595.

———. "Krakatoa." *Proceedings of the Royal Society of London*, May 1884.

———. "On the Volcanic Phenomena of the Eruption, and on the Nature and Distribution of the Ejected Materials." In *The Eruption of Krakatoa and Subsequent Phenomena*, ed. G. J. Symons, 1–56. London: Trübner, 1888.

Latter, J. H. "Tsunamis of Volcanic Origin: Summary of Causes, with Particular Reference to Krakatoa, 1883." *Bulletin Volcanologique*, 2d ser., 44, no. 3 (1981): 467–90.

Le Conte, John. "Atmospheric Waves from Krakatoa" [Letter to the editor]. *Science* 3, no. 71 (1884): 701–2.

Sandick, R. A. "Eruption on Krakatau." *Cosmos-Les Mondes* 8 (August 1884): 677–78.

Self, Stephen, and Michael R. Rampino. "The 1883 Eruption of Krakatau." *Nature* 294 (December 24/31, 1981): 699–704.

Thornton, Ian. *Krakatau: The Destruction and Reassembly of an Island Ecosystem.* Cambridge: Harvard University Press, 1996.

Van Doorn, M. C. "The Eruption of Krakatoa." *Nature* 29 (January 17, 1884): 268–69.

Verbeek, R.D.M. "The Krakatoa Eruption." *Nature* 30 (May 1, 1884): 10–15.

Zen, M. T. "Growth and State of Anak Krakatau Volcano." *Bulletin Volcanologique*, 2d ser., 34 (1970): 205–15.

Zen, M. T., and D. Hadikusumo. "Recent Changes in Anak Krakatau Volcano." *Bulletin Volcanologique*, 2d ser., 27 (1964): 259–68.

Related Reading

Furneaux, Rupert. *Krakatoa.* Englewood Cliffs, N.J.: Prentice-Hall, 1964.

8 • THE 1902 ERUPTION OF MOUNT PELÉE: A GEOLOGICAL CATASTROPHE WITH POLITICAL OVERTONES

Notes

1. Simkin and Siebert, *Volcanoes of the World*, 151.

2. Contour, *Saint-Pierre*, 2:127.

3. Ibid., 2:130.

4. Morris, *Volcano's Deadly Work*, 53.

5. Thomas and Witts, *Day the World Ended*, 129.

6. Ibid., 164.

7. Contour, *Saint-Pierre*, 2:137.

8. Morris, *Volcano's Deadly Work*, 55–57.

9. Ibid., 198.

10. Thomas and Witts, *Day the World Ended*, 293.

Cited References

Contour, Solange. *Saint-Pierre, Martinique*. Vol. 2, *La Catastrophe et ses suites* (in French). Paris: Editions Caribéennes, 1989.

Lacroix, Alfred. *La Montagne Pelée et ses eruptions* (in French). Paris: Masson, 1904.

Morris, Charles. *The Volcano's Deadly Work: From the Fall of Pompeii to the Destruction of St. Pierre*. W. E. Scull, 1902.

Simkin, Tom, and Lee Siebert. *Volcanoes of the World: A Regional Directory, Gazetteer, and Chronology of Volcanism during the Last 10,000 Years*. 2d ed. Tucson, Ariz.: Geoscience Press, 1994.

Thomas, Gordon, and Max Morgan Witts. *The Day the World Ended*. New York: Stein and Day, 1969.

Other Sources

Fisher, Richard V., Alan L. Smith, and M. John Roobol. "Destruction of St. Pierre, Martinique, by Ash-Cloud Surges, May 8 and 20, 1902." *Geology* 8 (October 1980): 472–76.

Heilprin, Angelo. *Mont Pelée and the Tragedy of Martinique*. Philadelphia: J. B. Lippincott, 1903.

Scarth, Alwyn. *Vulcan's Fury: Man against the Volcano*. New Haven, Conn.: Yale University Press, 1999.

Smith, Alan L., and M. John Roobol. *Mt. Pelée, Martinique: A Study of an Active Island-Arc Volcano*. Geological Survey of America Memoir, no. 175. Boulder, Colo.: Geological Survey of America, 1990.

Tanguy, Jean-Claude. "The 1902–1905 Eruptions of Montagne Pelée, Martinique: Anatomy and Retrospection." *Journal of Volcanology and Geothermal Research* 60 (1994): 87–107.

Related Reading

Fisher, Richard V., and Grant Heiken. "Mt. Pelée, Martinique; May 8 and 20, 1902, Pyroclastic Flows and Surges." *Journal of Volcanology and Geothermal Research* 13 (1980): 339–71.

Perret, Frank A. "The Eruption of Mt. Pelée, 1929–1932." *Carnegie Institute of Washington Publication* 458 (1935).

Tauriac, Michel. *La Catastrophe* (in French). Paris: France Loisirs, 1982.

Westercamp, D., and H. Traineau. "The Past 5,000 Years of Volcanic Activity at Mt. Pelée Martinique (F.W.I.): Implications for Assessment of Volcanic Hazards." *Journal of Volcanology and Geothermal Research* 17 (1983): 159–85.

9 • TRISTAN DA CUNHA IN 1961:
EXILE TO THE TWENTIETH CENTURY

Notes

1. Munch, *Crisis in Utopia*, 5.

2. Ibid., 86.

3. Ibid., 196.

4. Wheeler, "Death of an Island," 679.

5. Munch, *Crisis in Utopia*, 210.

6. Ibid., 228.

7. Ibid., 8.

Cited References

Munch, Peter A. *Crisis in Utopia: The Ordeal of Tristan da Cunha.* New York: Thomas Y. Crowell, 1971.

O'Connor, John M., and Anton P. le Roex. "South Atlantic Hotspot-Plume Systems: Distribution of Volcanism in Time and Space." *Earth and Planetary Science Letters* 113 (1992): 343–64.

Wheeler, P.J.F. "Death of an Island." *National Geographic* 121, no. 5 (1962): 678–95.

Other Sources

Baksi, A. K. "Reevaluation of Plate Motion Models Based on Hotspot Tracks in the Atlantic and Indian Oceans." *Journal of Geology* 107, no. 1 (1999): 13–26.

Booy, D. M. *Rock of Exile.* New York: Devin-Adair, 1958.

Brander, Jan. *Tristan da Cunha, 1506–1902.* London: George Allen & Unwin, 1940.

Helger, Patrick, and Michael Swales, comps. *Bibliography of Tristan da Cunha.* Oswestry, U.K.: Anthony Nelson, 1998.

Munch, Peter A. *The Song Tradition of Tristan da Cunha.* Bloomington, Ind.: Indiana University Research Center for the Language Sciences, 1970.

O'Connor, J. M., P. Stoffers, P. van den Bogaard, and M. McWilliams. "First Seamount Age Evidence for Significant Slower African Plate Motion Since 19 to 30 Ma." *Earth and Planetary Science Letters* 171, no. 4 (1999): 575–89.

10 · MOUNT ST. HELENS IN 1980:
CATASTROPHE IN THE CASCADES

Notes

1. Crandell and Mullineaux, "Appraising Volcanic Hazards of the Cascade Range," 10.
2. Crandell and Mullineaux, *Potential Hazards from Future Eruptions of Mount St. Helens*, C1-C2, C25.
3. Major et al., "Sediment Yield following Severe Volcanic Disturbance," 819–22.
4. Leik et al., *Under the Threat of Mt. St. Helens*.
5. Adams and Adams, "Mount St. Helens Ashfall," 252–63.
6. Budgen, "In a Giant's Wake," 46.
7. "Remember Spirit Lake," 2.
8. Shindler, "The Giants Are Only Asleep," 59–60.

Cited References

Adams, P. R., and G. R. Adams. "Mount St. Helens Ashfall: Evidence for Disaster Stress Reaction." *American Psychologist* 39, no. 3 (1984): 252–63.

Budgen, Mark. "In a Giant's Wake." *MacLean's* 93, no. 33 (1980): 46–47.

Crandell, Dwight R., and Donal R. Mullineaux. "Appraising Volcanic Hazards of the Cascade Range of the Northwestern United States." *Earthquake Information Bulletin* 6, no. 5 (1974): 3–10.

———. *Potential Hazards from Future Eruptions of Mount St. Helens Volcano, Washington.* Geology of Mount St. Helens Volcano, Washington, United States Geological Survey, bulletin 1383-C. Washington, D.C.: GPO, 1978.

Harnly, Caroline D., and David A. Tyckoson. *Mount St. Helens: An Annotated Bibliography.* Metuchen, N.J.: Scarecrow Press, 1984.

Leik, Robert K., Sheila A. Leik, Knut Ekker, and Gregory A. Gifford. *Under the Threat of Mt. St. Helens: A Study of Chronic Family Stress.* Washington, D.C.: Federal Emergency Management Agency, Contract No. FEMA/EMW-C-0454, February 1982.

Major, J. J., T. C. Pierson, R. L. Dinehart, and J. E. Coster. "Sediment Yield following Severe Volcanic Disturbance: A Two-Decade Perspective from Mount St. Helens." *Geology* 28, no. 9 (2000): 819–22.

"Remember Spirit Lake." *NSS News* 39, no. 1 (1981): 2.

Shindler, Tom. "The Giants Are Only Asleep." In *Washington Songs and Lore*, comp. Linda Allen. Spokane, Wash.: Melior Publications, 1988.

Tilling, R. *Eruptions of Mount St. Helens: Past, Present and Future.* U.S. Geological Survey, 1984.

Other Sources

Findley, Rowe. "Saint Helens: Mountain with a Death Wish." *National Geographic* 159, no. 1 (1981): 3–33.

Fisher, Richard V., Grant Heiken, and Jeffrey B. Hulen. "Mount St. Helens II: In the Path of Destruction." *National Geographic* 159, no. 1 (1981): 34–49.

———. "Mount St. Helens III: The Day the Sky Fell." *National Geographic* 159, no. 1 (1981): 50–65.

———. "Mount St. Helens Aftermath: The Mountain That Was—and Will Be." *National Geographic* 160, no. 6 (1981): 710–33.

———. *Volcanoes: Crucibles of Change.* Princeton, N.J.: Princeton University Press, 1997.

Peterson, Donald W. "Volcanic Hazards and Public Response." *Journal of Geophysical Research* 93, no. B5 (1988): 4161–70.

Stanley, W. D., S. Y. Johnson, A. I. Qamar, C. S. Weaver, and J. M. Williams. "Tectonics and Seismicity of the Southern Washington Cascade Range." *Bulletin of the Seismological Society of America* 86, no. 1 (1996): part A, 1–18.

Related Reading

Crandell, Dwight R., Donal R. Mullineaux, and Meyer Rubin. "Mount St. Helens Volcano: Recent and Future Behavior." *Science* 187, no. 4175 (1975): 438–41.

Lipman, Peter W., and Donal R. Mullineaux, eds. *The 1980 Eruptions of Mount St. Helens, Washington.* United States Geological Survey Professional Paper 1250. Washington, D.C.: U.S. Department of the Interior, 1981.

· Selected Bibliography

GENERAL

Blong, R. J. *Volcanic Hazards: A Sourcebook on the Effects of Eruptions.* New York: Academic Press, 1984.

Chester, David. *Volcanoes and Society.* London: Edward Arnold, 1993.

Decker, Robert, and Barbara Decker. *Volcanoes.* 3d ed. New York: W. H. Freeman, 1997.

Fisher, Richard V. *Out of the Crater: Chronicles of a Volcanologist.* Princeton, N.J.: Princeton University Press, 1999.

Fisher, Richard V., Grant Heiken, and Jeffrey B. Hulen. *Volcanoes: Crucibles of Change.* Princeton, N.J.: Princeton University Press, 1997.

Francis, Peter. *Volcanoes: A Planetary Perspective.* Oxford: Clarendon Press and Oxford University Press, 1993.

Ollier, Cliff. *Volcanoes.* Oxford: Basil Blackwell, 1988.

Scarth, Alwyn. *Vulcan's Fury: Man against the Volcano.* New Haven, Conn.: Yale University Press, 1999.

Sigurdsson, Haraldur. *Melting the Earth: The History of Ideas on Volcanic Eruptions.* Oxford: Oxford University Press, 1999.

Sigurdsson, Haraldur, Bruce F. Houghton, Stephen R. McNutt, John Stix, and Haxel Rymer, eds. *Encyclopedia of Volcanoes.* New York: Academic Press, 2000.

Simkin, Tom, and Lee Siebert. *Volcanoes of the World: A Regional Directory, Gazetteer, and Chronology of Volcanism during the Last 10,000 Years.* 2d ed. Tucson, Ariz.: Geoscience Press, 1994.

Tazieff, Haroun. *Craters of Fire.* Trans. Eithne Wilkins. New York: Harper, 1952.

FICTIONAL BOOKS RELATED TO VOLCANOES

Bulwer-Lytton, Sir Edward. *The Last Days of Pompeii.* Garden City, N.Y.: International Collectors Library, 1946.

Cameron, Ian. *The White Ship: A Novel of Adventure.* New York: Charles Scribner's Sons, 1975.

Catto, Max. *The Devil at Four O'Clock.* New York: William Morrow, 1959.

Daniels, J. R. *Firegold.* New York: Coward, McCann and Geoghegan, 1975.

Du Bois, William Pène. *The Twenty-One Balloons.* New York: Dell, 1969.

Hayes, Joseph. *Island on Fire.* New York: Grosset & Dunlap, 1979.

Innes, Hammond. *Angry Mountain.* New York: Harper and Brothers, 1950.

Lloyd, Alan. *Alive in the Last Days of Pompeii.* Los Angeles: Pinnacle Books, 1975.

Malaparte, Curzio. *The Skin.* Boston: Houghton Mifflin, 1952.

McKenney, Kenneth. *The Fire Cloud.* New York: Simon and Schuster, 1979.

Sontag, Susan. *The Volcano Lover: A Romance.* New York: Farrar Straus Giroux, 1992.

Warner, Sylvia Townsend. *Mr. Fortune's Maggot.* In *Four in Hand: A Quartet of Novels,* 135–263. New York: W. W. Norton, 1986.